Learning Science

For Peter Fensham and Bob Gagné
with thanks for many years of
friendship and encouragement

Learning Science

Richard T. White

Basil Blackwell

First published 1988

Basil Blackwell Ltd
108 Cowley Road, Oxford, OX4 1JF, UK

Basil Blackwell Inc.
432 Park Avenue South, Suite 1503
New York, NY 10016, USA

British Library Cataloguing in Publication Data

White, Richard T.
Learning science.
1. Science – Study techniques
I. Title
507 Q181
ISBN 0-631-15698-4
ISBN 0-631-15699-2 Pbk

Library of Congress Cataloging in Publication Data

White, Richard T. (Richard Thomas)
Learning Science.
Includes index.
1. Science – Study and teaching. 2. Learning, Psychology of. I. Title.
Q181.W48 1988 507'.1 87-18228
ISBN 0-631-15698-4
ISBN 0-631-15699-2 (pbk.)

Typeset in Garamond 10.5 on 12 pt
by Joshua Associates Ltd, Oxford
Printed in Great Britain by
Page Bros Ltd, Norwich

Contents

Acknowledgements

Much of this book was written during a period of leave given by Monash University, which I spent at the University of Leeds. I am grateful for the hospitality and friendship I received in Leeds. The British Council supported my travel with a generous grant. Mrs Cath Henderson and Mr Claude Sironi provided valued and much appreciated assistance in the production of the manuscript.

Explanatory Note

Although schools do not differ all that much from one country to another, some of the terms that are used in describing them do vary. Among these is the way of referring to year levels. In America and Australia the common practice is to speak of grades, from 1 to 12, whereas in Britain grades are used for primary (or elementary) school after which the number system starts again for forms in secondary school. The grade system is used in this book; readers used to the British system may make an appropriate conversion for grades above 6 by subtracting 5 and calling them forms. Grade 12 refers to the final year of secondary schooling, the upper sixth form in Britain.

Preface

My theme is that learning is an active, not a purely receptive, process in which people construct their own meanings for new information. This process of construction involves characteristics of the person, such as existing knowledge, abilities and attitudes, which have their roots in experiences and in genetic inheritance, and the context in which the learning occurs.

All animals learn, but only in humans has learning for more than immediate survival become a central part of existence. We learn about many things: our language, our history, our social systems. This book is about our learning of science; about how we interpret and acquire information about the universe in which we find ourselves. The scope includes, but is not limited to, formal schooling, for an important development in recent thinking about the learning of science is the recognition that people form views of the world from experiences outside the classroom, and that this learning may be more permanent and influential in their lives than anything they are told in school. Nor does the book only address learning by a restricted age group. Learning does not occur in stages, nor at only one period, but throughout life. It is an accident of social history that we concentrate formal learning of science in the second decade of life, and it is one of the tragedies of this century that for many people learning almost stops before they are 20 years old. Avoidance of that tragedy is a theme that recurs throughout the book. The psychology described here applies to all ages.

All books are biased: inevitably they reflect the writer's view, which in the present instance may be labelled 'constructivist'. That is, the view that each individual constructs his or her own picture of the world. Of course people are influenced in this by the information they receive from others. The way that information is presented, the terms that other people use and link together, the things that are pointed out as important and the interpretations others give to shared experiences, all affect each individual's construction of reality. Consequently, there are similarities in people's views. Nevertheless, the picture of the world is the learner's own, and it will differ from the pictures of others. Aristotle and Galileo saw different relations between force and movement; modern scientists, in contrast to their predecessors, do not think of heat as a fluid which can be expressed from objects; rainbows are now interpreted as reflections and refractions in water drops rather than as reaffirmations of God's promise never to send another flood. These differences across ages are paralleled by differences between people of the present. I see waves as the consequence of the force of the wind on initially slight irregularities in the water surface, while children have told me they are caused only by passage of ships far out to sea. Other examples, from recent investigations of differences between children's and teacher's or scientists' views, are described in chapter 5.

The universe impinges on us, and we are aware of it through our senses. From the sensations we receive we construct objects and incidents and determine causes for effects. Each of us builds a world. The book describes a model of how this is done. Psychologies of learning, like other sciences, are models. That is, they are descriptions, expressed in terms of invented concepts, which attempt to explain and predict phenomena and so confer greater control over outcomes. Since the concepts are invented and, as we will consider later, mean different things to different people, models are constructions. Not only are they constructions themselves, they influence further constructions – the ways we interpret the world.

To give an example, consider the very basic concept of time, which is spoken of in most cultures as if it were a regular, continuous progression in one direction along one dimension. It is represented mathematically in that way in our science, and we are so imbued with it that it is hard to appreciate that there are other ways of looking at it. The Hopi Indians experience time just as we do, of course, but think and speak about it in a very different way. 'Among the peculiar properties of Hopi time are that it varies with each observer, does not permit of simultaneity, and has zero dimensions; i.e., it cannot be given a number greater than one. The Hopi do not say, "I stayed five days", but "I left on the fifth day" ' (Whorf, 1940, p. 216). Whorf has called Hopi a timeless language. He points out that although the Hopi can describe the universe quite adequately, their non-dimensional view of time means that they do not share our concepts of velocity and acceleration. They are not wrong in their view, nor are we; it is just that our constructions differ. Even within cultures that follow modern science, time may be seen differently. Mori, Kitagawa and Tadang (1974) demonstrated that religious beliefs affected Thai and Japanese conceptions of time. The

Buddhist Thais tended to think of time as a circular succession, infinitely recurring with no beginning or end; Japanese from Christian schools more often thought of it as linear, with a beginning and an end; and Japanese from public schools tended to think of it as linear with a definite beginning but infinite in extent.

The psychology of learning I present in this book is a model, like any other. Naturally it has been influenced by other models. Among those I have been most conscious of in developing it are classical and operant conditioning, information processing, and the models of Gagné, Ausubel, Piaget, Tulving, Schank and Abelson, Paivio, Wittrock, Carroll and Marton. Something has been absorbed from each of these and others, and interpreted in the light of my experiences to form a new synthesis, one that fits my own purposes. Most of my sources discuss learning in general and in a context-free way. My model is more specific, directed at the learning of science in formal instructional and everyday contexts. It could, however, be a useful model for other learning processes, and indeed I will often draw examples from other subjects. It was, though, developed with science in mind.

Bruner (1966) distinguishes between theories of learning and theories of instruction. The former are descriptive, the latter prescriptive, concerned with improving rather than describing learning. Like many artificial dichotomies, this division is not as sharp as it appears. Just as Piaget's model of development has implications for learning and for instruction, so has my model of learning for instruction and development. Development, learning and instruction are so inter-related a set of concepts that anything said about one must bear on the others.

Bruner stated four major features of a theory of instruction. It should specify:

1 Experiences which predispose people to learn.
2 How a body of knowledge should be structured so it is most readily learned.
3 Effective sequences of presentation.
4 The nature and spacing of rewards and punishments.

Although I do not address these issues directly, the propositions that form my theory do bear on them. Implications of the model for teaching, the curriculum and the organization of schooling are referred to throughout the book, and are collected together in the final chapter. In that way the book is prescriptive, as my aim is to *improve* the learning of science, not merely to describe it. It is, however, a book about learning rather than teaching, for it does not try to cover everything that teachers do. There is little in it about discipline and class control, though a theory of learning is not totally irrelevant to that since one source of conflict is students' frustration over their learning. Better learning and better behaviour interact, each promoting the other. However, though teaching is referred to frequently, it is always in regard to acts that directly foster better learning. Learning of science is the focus of the book.

1

Values and Aims

After eight months of experience in the Project to Enhance Effective Learning (Baird and Mitchell, 1986), two tenth-grade students came to their science teacher.

'We see what all this is about now,' one said. 'You are trying to get us to think and learn for ourselves.'

'Yes, yes,' replied the teacher, heartened by this long-delayed breakthrough, 'that's it exactly.'

'Well,' said the student, 'we don't want to do that.'

This incident illustrates what every experienced teacher knows, that values are crucial in determining both what is taught and what is learned. Psychology cannot be divorced from values. 'Learning of science' is a broad phrase, which leaves open exactly what it is that is to be learned. People may differ not only on the fine detail of what content they think matters, but fundamentally on the purpose of learning. The aims that people express for learning, for courses of study and for institutions, reflect their values. Aims are specific ways of giving force to values. At the outset, then, I should set out at least a general statement of the values I hold that are relevant to the learning of science.

The observation that values are not absolutes, not truths to be discovered or revealed, and that they are arbitrary, determined by individuals and societies for themselves and occasionally to impose on others, is in itself a reflection of my values. In making it I display myself as a creature of my times, for there have been many periods in the history of the world when such an observation would not have been made, or if made, not accepted. Even today many would dispute it, holding that values are revealed, fixed by God. Values are nearly always in dispute, though at certain times the argument is more heated than at others.

Values differ between communities. Where I live the rights of the individual are valued more in comparison with the rights of the State than in most other countries. The people around me believe in progress and development, while the Australian aborigines value stability, conservation, continuity; we think that land belongs to us, but they that they belong to the land. Values change within communities. Democracy is valued, though in the last century democracy was a pejorative term in Western societies. Some values are stable for generations: the desirability of peace, social order. We can see others changing: the function of work, the roles of men and women.

Like other values, those for education change with place and time. Indeed, they are more changeable than most, since they are values that are held about what should happen for someone else – usually children, one's own and others – and apply only tangentially to one's self. When people debate the purposes of education, they mean the education of others, rarely themselves. Presumably they assume that, for them as adults, education is complete. That is itself a value, to which we will return. We are freer in deciding what should be done by others because we are unhampered by concomitant costs in personal freedom and other inconveniences.

As well as applying to others, values in education are derivative, in being subservient to more general societal values: religious convictions, including humanism and political faiths; views of the functions of the Church, State and family; and national aspirations. They are open to the influences of economic and political forces. They vary with the personalities and experiences of individuals.

Examples of these influences are easy to come by, especially for national aspirations. The paradigm is Nazi Germany, where as Kneller (1941) has made clear, the aim of the system was to promote a conformism and loyalty to the State and its leader which in our value system appear excessive. Conformism is behind all education, of course. Even where the value is for individual development and self-expression, the aim is to have people conform to our notion of what is good. The great science curriculum movement of the 1950s and 1960s – PSSC (Physical Science Study Committee), Chem Study, BSCS (Biological Sciences Curriculum Study) and so on – stemmed from a national aspiration. Science was seen as the key to maintaining or recovering the pre-eminence of the United States, so science education was valued and supported vigorously. Some echo of that can be seen now with computer education, in several countries. The study of computers is advocated not for the cultural well-being of individuals, but for the sake of each nation's economic future. The same motive may lie behind the acceptance in Thailand of the value Western nations put upon science education and the ways in which it is taught. Thailand has strong traditions, and values derived largely from Buddhism. The Thais have always been independent, so their values were never disturbed by a colonizing power. Until recently teaching methods in Thai schools were markedly different from those in the West. Teachers would rarely ask questions of the class, and never of individuals; learning was totally receptive. Since 1970 there has been a determined effort to change to Western-style curricula and methods in science and mathematics education. Values have shifted from maintenance of a tradition to a belief that science is the root of Western economic power, and that to maintain other vital Thai values such as independence the education system must change.

Western systems are under pressure from shifts in values across time. A theme that recurs in this book is that the form of organization of schools was settled in the nineteenth century and is now out of line with our values. In the nineteenth century values in most countries were such that secondary education was not seen as a responsibility of the State. In England the public schools existed to produce leaders of society from the upper middle class males and so ensure the continuation of an elite. An enjoyable way of recapturing the values of that time is to read schoolboy novels such as Talbot Baines Reed's *The Fifth Form at St Dominic's*.

Even in a society as egalitarian as Australia, for the first half of the twentieth century secondary schools aimed at producing an intellectual elite – a meritocracy, and a conformist one. It was also a society that valued education for males above education for females. It was not until the mid 1970s that as many girls as boys stayed on till the final years of secondary school. I recall a personal experience from 1962 illustrating this difference in values for boys and girls. Late in her tenth grade, a girl at the school at which I then taught was in distress because her parents wanted her to leave. The parents believed that girls did not benefit from too much schooling, because they were not going to need it for a lifetime career. In the end the teachers' and the girl's values prevailed over the parents', and the girl continued at school. The values of the parents in this case appear strange now, but were quite common 20 years and more ago.

Most nations have come to put greater value on more education for all, though our systems are still trying to find out how to cope with it. At the same time, and perhaps for the same economic causes, the emphasis has shifted from a narrow conformism to the development of the individual. Values are often expressed in catch phrases, and 'Let a thousand flowers bloom' and 'Do your own thing' were popular ones in the 1960s. One can see attempts by schools to cope with that shift in values – freer relations between teachers and students, attention to pastoral care, weakening of rules for behaviour, greater choice of programs and subjects. Individual freedom may have reached its height as a value at the end of the 1960s, when economic prosperity peaked. Pressure to conform may come again as values shift with economic recession or worsening of international relations.

Most of the values that I described in the previous section concerned groups, whole societies. The values of societies, however, are no more than the views that are held by the great majority of their members. Although values really belong to individuals, in education as in other large social enterprises it is the general view that determines what happens. Sometimes individuals can put their values into operation in the education of their own or a small number of other people's children – a notable example is A. S. Neill with his Summerhill school (Neill, 1937) – but generally education is a large and formal system involving many people, none of whom has total power to impose his or her values on the rest. Values within the group will vary, but there must be a core of accepted values without which the group would cease to exist. It would fragment, and the parts would go their own ways or resort to force to coerce the others into their way of looking at things. Often the cohesive values are not overt, or are rarely discussed. Recognized or not, they are the source of the aims of the system.

Aims are debated, by philosophers, educationists, administrators, politicians, teachers, parents, subject specialists and children. All have characteristic ways of influencing aims. Children, for instance, do not make the direct assertions of philosophers and other educationists, but affect aims by their reactions to procedures. For example, the aim might be to produce an intellectually oriented, cultured community, with knowledge of science, mathematics, history, literature, languages and art. This will take time to transmit, so school attendance is made compulsory until the age of 16. Large numbers of children may then refuse to cooperate, misbehaving or playing truant. When the system is perceived to be failing, the authorities may try new procedures, at first simple ones like school police or high fences to keep students in. When these, and more enlightened and subtle attempts fail, the aim is abandoned.

Unattainable aims are discarded, even if values imply they are important. Psychology influences the choice of aims by detailing how some may be attained and by indicating which will prove difficult. Earlier I asserted that the aims we are interested in determine the psychology we need. Now it is seen that the relation is reciprocal: the psychology indicates which aims may be attempted and which might as well be abandoned. Psychologists are more than facilitators: they join children, philosophers, administrators and the rest as contributors to the aims of the educational system.

The contributors to aims may conflict. Some may see education as a tool for bringing about social changes, others for maintaining the status quo. Within a group with the same overall aim, for example promoting a scientifically informed population, there may be differences over more specific aims that combine to form the overall one, such as whether the population should acquire subject matter (product) or the methods of scientific thinking (process). These conflicts usually lead not to a static consensus but rather a dynamic equilibrium, where debate continues between proponents of differing views. Examples are State aid to non-government schools in Australia and the United States, tripartite versus comprehensive systems in Britain, and the adoption of a curriculum package anywhere.

Because so many people and such diverse groups are involved, the determination of aims of education (or a part of it, such as science education) is a complex process. Forming aims can take a long time. Consequently aims are often unstated, unclear, forgotten, and slow to change even though values may have shifted long before. Between periods of social ferment, when values are changing fast and aims are stated and argued over, there are calms when aims are taken for granted, not set out in any but a ritualistic manner, and not thought about or debated. Then the education system can become automatic, running on from year to year in the same way that people recall it always has done. Issues of aims, the nature of the curriculum, the purpose of each subject, are not prominent; the schools jog along quite well. In many ways that is a comfortable experience. There is no uncertainty, nearly everyone knows what they should be doing, even if not why. James Hilton's novel *Goodbye Mr. Chips* (1934) conveys the atmosphere of such times. Chips, the lovable old schoolmaster, was able to tell the same jokes to grandsons of his earlier pupils, in the course of almost identical lessons. He also resented the slightest change to his settled ways.

In stable times newcomers are readily inducted into the practices at which the old hands are already skilled. If the community's values really are stable and have not yet got out of tune with the practices of its education system, the newcomers are unlikely to question why things are done in such a way. If they do, however, they will most likely be told 'We have always done it that way'. Persistent questioning could lead to exclusion from the group running the system.

Unstable times strain educational systems. They are so large, and have so many groups contributing to their aims, that it is inevitable that in dynamic times their practices will lag behind shifts in values. Gradually they cease to serve the real needs of the community.

Newspapers from periods such as the 1880s, which now appear to have been quiet and undisturbed, show that people generally think they are living in dramatic times. There are, however, solid grounds for judging our present era as unstable, and so we might expect to observe discrepancies between our values and practices. Apart from the depression of the 1930s, the first three-quarters of the twentieth century were years of increasing wealth and comfort, both in total and in the spread in most countries to a greater proportion of the population. At the same time as the population grew, rapid developments in technology led to widespread replacement of manual labour with machines. In the most recent phase of this revolution, the proliferation of computers and associated tools such as robots and word processors, an even wider range of jobs is becoming redundant. There is now less need for labour, greater need for skilled operators and more leisure time. The contrast in personal freedom between the present and the start of the century is brought out sharply in social histories such as Blythe's *Akenfield* (1969), which describes how in the course of a lifetime conditions of grinding poverty and unremitting physical effort vanished from the English countryside. Technology has also made possible fearful weapons, and people now have to live with the possibility that all life could be rapidly destroyed. Although many now enjoy greater comforts than during the first decade of this century, ours is a tense age. To read Siegfried Sassoon's *Memoirs of a Fox-Hunting Man* (1928) is to enter a period before the First World War of calm and settled order that is in poignant contrast with today.

Shifts in values accompanied these economic and technological developments. Colonialism was widely accepted at the beginning of the century; by the middle years it was widely rejected. Social class, hierarchies of status, used to be far more valued: by 1950 most of the monarchies of 1900 had been swept away. Some of the most visible clashes between values occurred during the height of prosperity in the late 1960s, towards the end of the Vietnam War, when unrest spread through the West and there were bitter confrontations. Although there were many elements underlying the ferment of the late 1960s, rejection of authority was a marked feature. In some places – Paris, for example, where for a while students disputed control of the streets with armed police – this reached such a pitch that it appeared that revolution could occur.

Throughout this turmoil, education systems that had been established to meet the aims of basic literacy and numeracy for all and the production of an elite meritocracy became less and less in tune with the new values that were emerging. At the beginning of the century all children in the more prosperous countries attended primary school, but only a minority had secondary schooling and a tiny fraction went on to tertiary education. This fitted the values of the time quite well. Indeed the very compression of formal schooling into the early years of life is a reflection of a stable period. In unstable times people are less likely to believe that you can launch a person at age nine, or 15, or 20, fully equipped to handle the next 60 years. Now the new value is education for all for as long as possible. Practical attempts to continue it beyond adolescence include the community schools of Britain and other countries (Poster, 1982) and the Council for Adult Education in Australia. As yet these do not touch many people, but at least all children in advanced countries go to secondary school for several years, and substantial proportions reach tertiary studies.

Although the proportion attending secondary schools is much greater now than early in the century, what happens in those schools has changed little, and so these schools are now in tension with their society. The strains are not so severe in primary schools, because they were established to handle mass education. Tertiary colleges are in less difficulty than secondary schools because, although they have had to accept much greater numbers, they have not yet had to cope with education for all.

The stresses in secondary schools are visible in general matters as well as in the specific case of learning in science. One example is the restriction of personal freedom. Behaviour is much more circumscribed than in homes and in public places outside schools. In many countries schools have requirements for dress and personal appearance which reflect the values of generations before. A further example is the level of comfort in schools. Furniture, floor coverings, general appearance are more in accord with general standards of the turn of the century than in any other institution except perhaps prisons. In 1900 schools were no less comfortable than most offices and the general level of homes. They have not accommodated to the shift in standards of luxury. This also applies to primary schools.

A more subtle way in which secondary schools have been left behind by changes in values is in the maintenance of school subjects. The division of the day into subject areas, each the responsibility of a different teacher, reflects values. It marks the acceptance of the main function of secondary schooling as the transmission of certain sorts of knowledge. It reflects a view of knowledge that is limited to facts and algorithms, and stands in the way of attainment of other forms of learning which are now necessary. It may seem odd in a book on the learning of a particular branch of knowledge to suggest that the curriculum should not be a set of formal disciplines, but, as I will try to make clear, my objection is not to disciplines but to the attempt to transmit their full meaning through a system which separates them from each other and, apparently inevitably, from the experience of life outside formal schooling. In the meantime it would be unrealistic not to take account of the system as it is. The psychology set out in this book has implications for change in common practice, but is intended to be useful even within the present system.

Once it is accepted that the curriculum is to consist of subjects, one has to consider whether the subjects included are the best mix for meeting the aims a society has for its schools. Just as the general form of organization of education may lag behind changes in aims, so at another level the choice of subjects may not be the most appropriate possible. Classics once made up most of the curriculum because it was seen as a means of transmitting the wisdom and philosophy of a golden age which the receivers would find valuable in their lives as leaders of society. The common sense psychology of the time may also have used a muscular analogy, that mental exercise in construing blocks of Latin or Greek and in committing them to memory strengthened mental facilities. Even when this rationale was abandoned, classics hung on in the curriculum. This may have been due in part to the complexity of factors determining the nature of the educational system, which I have discussed already, in part to another influence which retards deletion of old subjects and introduction of new, the vested interest of teachers. Teachers can have careers of 40 years or more. They are trained in a particular discipline and with experience develop confidence in their ability to teach it. They think of themselves as teachers of science or mathematics, French, history, art. They will try to thwart attempts to remove their speciality from the curriculum, even when others demonstrate it is serving no useful purpose: it is their livelihood, their life. On the other hand, attempts to introduce new subjects may fail because no specialists are available. Thus French continues to be taught in most secondary schools in Australia because there are lots of teachers of French, while Japanese, for which a case could be made as a more relevant language for the nation's present and future interests, is not widely available because of a lack of people who could teach it.

Other examples of the lag between need for and introduction of a subject are law and economics. One could argue that appreciation of both should be an aim for the education of all citizens, but these subjects have been slow in gaining a substantial share of the curriculum. Computer studies, which are spreading fast, may be an exception to this lag principle.

Science, on the other hand, already has a place in the curriculum, and might be thought to be secure; but there are threats to it. One is pressure from new subjects, such as computer studies, elbowing for places. Of itself, that would not be sufficient to displace science, but it is allied with two practical concerns – the cost of science and the shortage of science teachers. The laboratories, apparatus and consumables required by science make it an expensive subject. Cost-conscious administrators therefore see it as a prime target for economies. The new design of government schools in Victoria does not include laboratories as an integral part of the building. Science teachers are chronically in short supply throughout the world. When the economy booms they find places in industry; during the 1960s the median number of years' experience of physics teachers in Australia fell as low as two, as vigorous attempts by administrators to have more trained were countered by a fearsome resignation rate. Even in the 1980s, a period of recession and a general over-supply of teachers, there are not enough teachers of science, especially physical science. Factors such as these could lead to the disappearance of science from the curriculum. The defence against these practical threats must be that science enables fulfilment of aims that are important to the community.

What functions then does science serve? The answers determine the range of learning to be included under the umbrella name 'science', and how this learning should be promoted, that is, how science should be taught. Before discussing what the aims of science might be, I would observe that understanding of science may be most needed in unstable times like the present, because it assists people to cope with the rapid introduction of new technologies and equips them to make informed decisions about new situations for which routine procedures are not yet available.

Science must be seen within the wider context of the functions of all learning. In classifying views of the school curriculum, Eisner and Vallance (1974) identify four functions: the promotion of skills which enable people to learn anything, irrespective of content; the development of self, a personal integration achieved through satisfying experiences which relate to life outside school; acceleration of change in the values and procedures of society; and the transmission of knowledge through established disciplines, to enable the recipients to participate fully in their culture. The balance one requires between these is a personal choice. Mason (1970), for instance, gives first priority to promoting self-esteem. This does not eliminate transmission of knowledge, since Mason argues that self-esteem includes seeing all human learning as one's heritage, a playground for lifelong enjoyment. Presumably to benefit from this heritage people must have skills of learning, too. Berman (1968) expresses essentially the same view as Mason, listing processes of loving, knowing, creating and others that are involved in living as an adequate contributor to the world. Integrity of self is bound into this ideal of service.

The justification for the learning of science must be that, wherever the balance comes in these broad aims, it is capable of a unique contribution. For self-esteem, or self-actualization, the case could be that these desired outcomes are tied into a sense of control over one's world, for which science is vital. Control implies understanding of causes and effects, being able to predict what consequences will follow from an act, being able to explain why something happened. Science does provide that form of understanding for much of what we experience. Not all, however: Munby and Russell (1983) point out that science is not the only system of rational explanation which is important in understanding ourselves and our worlds. History and other disciplines have their rational methods also.

Wherever the balance comes between the approaches identified by Eisner and Vallance, the designers of science curricula have to produce practical ways of attaining general aims. They do this by planning to give students knowledge of some subject matter, facts and principles about the natural world; some acquaintance with the processes of science, how scientific knowledge is derived or constructed; some skills useful in solving problems; and some general ways of thought. Designers vary in which of these things they choose as important. In the PSSC course (Physical Science Study Committee, 1965), for instance, there is less emphasis on the practical application of physics in society than there was in older courses, and more on understanding of broad conceptual schemes such as the function of models in physical science, principles of measurement, and the pervasive importance of the concept of energy. The later *Project Physics Course* (1970) put more emphasis on the historical roots of scientific ideas. Variations in emphasis are not surprising, given that people from many backgrounds contribute to aims. Nor is the degree of common ground surprising, if there are generally accepted values from which detailed aims derive.

A case can be made that learning of science does contribute to values that today are widely held. Aims can be derived from these values and incorporated in the design of science courses. However, we need to appreciate that the aims that are stated may not be identical with the aims that are attempted. Whatever curriculum writers, politicians, administrators and academics say are the aims of a course, what actually happens is in the hands of the teachers and students. The syllabus may advocate the development of habits of critical evaluation, but if the teacher sees that as irrelevant compared with the learning of facts it will not be achieved. An even more crucial element is the aim of the students. If they perceive that their interests are best served by passing an examination, then all aims that do not contribute to that goal will be ignored. The students' selection of aims will in turn affect the learning strategies they develop.

Aims of students, aims of teachers and aims of curriculum designers interact to determine what is learned and, which is not the same thing, what is taught. They also influence *how* it is taught.

Another factor that influences what is taught, and how, is the psychology of learning which underlies the decisions of curriculum designers and teachers. If learning is seen as the simple accretion of facts, then the syllabus will consist of an encyclopedic collection of facts. This is demonstrated by old texts (see figure 1.1) and, unfortunately to my mind, many modern ones also. Those facts will be taught in a simple way: the teacher will tell them to the students, who will be drilled in their recall. Tests, as reviews of standardized ones from as late as the 1950s reveal, will be measures of whether the facts can be recalled (see Buros, 1953, for reviews). If, however,

The Mechanical Powers.

A machine is an instrument by means of which force is applied to the performance of work, generally by changing the direction of the force—as the capstan, by which sailors raise the anchor; the crane, by which stones or other heavy weights are raised; or the lever, by which a heavy object is moved *upwards* by pressing *down* the other end. Such contrivances do not increase the force applied, but merely afford the means by which it may be directed more advantageously to the end in view. The simple machines used for this purpose, or the MECHANICAL POWERS, as they have been called, are usually considered to be six in number—the Lever, the Wheel and Axle, the Pulley, the Inclined Plane, the Screw, and the Wedge.

I.—The Lever.

A lever, from Latin *levo*, to raise, means *that which raises* or *lifts*. It is a body of any form fixed at a point about which it can move, called *the centre of motion*, or the *fulcrum* [Latin, 'a prop']; thus, when the fire is stirred with a poker, the poker is, for the time being, a lever, and the bar on which it rests is the centre of motion, or fulcrum. Levers are distinguished into three classes.

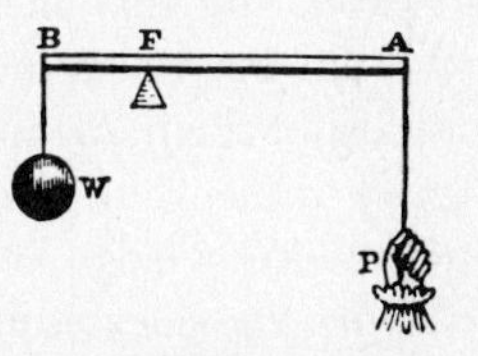

Fig. 2.
AB, the lever; F, the fulcrum; W, the weight; P, the power or force.

(1) The first class of lever (fig. 2) has the fulcrum between the force applied and the weight to be raised, or the resistance to be overcome, as in the case of a poker when stirring the fire. The common balance and a pair of scissors—the latter being double—are both examples of this class of lever.

(2) The second class of lever has the weight between the power and the fulcrum (fig. 3). The common wheelbarrow is a lever of this class; an oar is another example—the man pulling being the power, the water taken hold of by the blade being the fulcrum, and the boat the resistance.

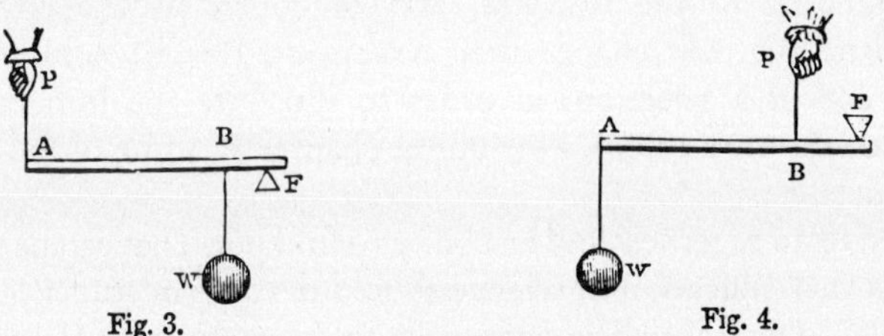

Fig. 3. Fig. 4.

(3) The third class of lever has the power between the fulcrum and

Figure 1.1 A page from a nineteenth-century text (from Chambers and Chambers, 1869)

learning is seen as a more complicated matter, the details of the syllabus and the methods used to attain the aims will be very different.

As I argued earlier, in producing a psychology of learning of science I must have formed a view of the purpose of the learning. The aims that animate this book are partly those that now are widely accepted, for otherwise the book would be a phantasm, dealing with a non-existent world. Associated with these current aims are others that are not generally valued, but which I think matter. Their presence indicates that my intention in writing the book is to bring about changes in the learning and teaching of science.

My values for education in science are that it should assist all to maintain an eagerness and an ability to find out all that they can about their world, throughout life; that in consequence people will comprehend their environment and want to improve or preserve it wherever that will contribute better to the comfort, both physical and aesthetic, of themselves and future ages; that their beliefs about the universe should be in accord with observation, and that they will see it as something to be understood in rational, not magical terms; and that science will be appreciated as one of the noblest of the humanities.

Clearly these values require that one aim is that people acquire an amount of knowledge of scientific facts and principles, an aim that is shared with the encyclopedic courses of the past. Comprehension of the environment, however, requires more than acquisition of facts as an aim: there must be understanding as well. Understanding is also involved in seeing the phenomena of the universe as rationally explicable. Describing what is involved in understanding is part of the psychology of learning that is set out in this book. The selection of appropriate content is subjective, since people may see different aspects of knowledge as relevant to understanding of the universe, and of course as the environment changes so people may change their selections. Despite subjectivity and the transience of a selection, in order to illustrate the implication this value has for the curriculum, I set out in chapter 10 some of the content I see as important.

I referred to an eagerness and an ability to find out. That can be translated into an aim that education in science should develop in students responsibility for their own learning and strategies that are applied in the acquisition and comprehension of knowledge. The psychology of learning must describe how that aim can be achieved.

The desire to improve or preserve the environment refers to a tendency to behave in a certain way. Traditionally, tendency to behave is one of the aspects of the human characteristics that are known as 'attitudes'. Therefore, the psychology must describe the relation between attitudes and other things that are learned, and must give a hint as to how they can be formed or manipulated.

In sum, the psychology of learning that is set out in the succeeding chapters is addressed to the development of independent, reflective learners, who have acquired and comprehend sufficient knowledge of the natural and technological world for them to interpret their surroundings and control them in ways that benefit themselves and others.

2

Factors Influencing Learning

When I ask teachers what determines how well students learn science, they usually mention the quality of the teaching and characteristics of the students, such as their abilities, how much they know about the subject to start with, and their willingness to learn. In identifying these factors teachers are expressing a theory of learning, a theory that has been formed from experience and that is passed on from one generation of teachers to the next through conversation about individual students and classes.

The theory exists for most teachers as a number of near-axioms, which they do not need to see set out as formal statements. However, for teaching to shift from a craft to a profession, vague unconnected postulates must be developed into a coherent and detailed theory. The purpose of this book is to speed that development by presenting descriptions and predictions about the acquisition of knowledge of science and its application in understanding natural phenomena. The descriptions are explanations of meanings of concepts, and the predictions are assertions about the qualitative or quantitative values that concepts will take under given conditions. An example of a descriptive proposition from physics is 'Kinetic energy is the energy a mass has by virtue of its motion'. A predictive proposition from biology is 'When a population is put under stress, its variation will decrease'. Parallel examples for this psychology of learning are: (descriptive) 'A person's concept of energy is the total set of propositions, strings, images, episodes, and intellectual skills that the person associates with the label "energy"'; and (predictive) 'students who are taught for a long time by transmissive teachers will develop learning strategies for attention but not for reflection'.

In their early stages disciplines are marked by lack of consensus over descriptive propositions and by theories that have a relatively small proportion of predictive propositions. Although education has been practised and thought about for as long as humans have existed, as a systematic field of study it is young and so the theory I put forward here contains much description. Wherever possible, though, I have made predictions. It would be irresponsible not to, for although descriptive propositions may be overthrown or at least discarded, the test of a theory is whether its predictions are fulfilled. The predictive propositions are the targets of research. Investigations can lead to clarifications of concepts, that is to amendments of the descriptive propositions, but their main function is to check the accuracy of the theory through the truth of its predictions.

Correct predictions are all that is required of a theory in the pure sciences. Education, however, like medicine, law and engineering, is an applied subject. It is practised in a social context, so another test of an educational theory, besides the accuracy of its predictions, is its utility. Does it, or has it the potential to, affect practice? I will try, by means of examples throughout the chapters that follow, to show how the ideas in the theory can be applied in teaching.

This attempt to convert to detail teachers' common views of learning begins with the pattern shown in figure 2.1 of interactions between major concepts. Some of the concepts – knowledge, abilities, attitudes, physical state and needs – refer to the state of the learner; some are the influences that determine those states – genetic inheritance and experiences; others are the circumstances in which the learner is placed – context and instruction; and what the learner makes of them – perception of context. All lead to what the individual does, in the form of learning or an overt act.

The arrows in figure 2.1 represent causal influences. Thus learning is shown as a performance which is determined not just by teaching but also by the learner's knowledge, ability, attitudes, needs and perception of the

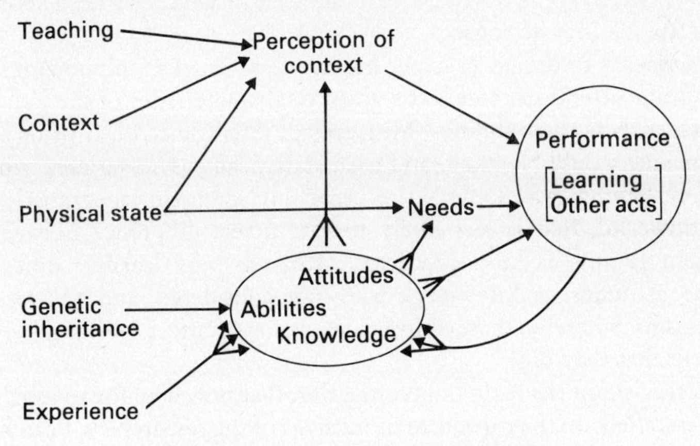

Figure 2.1 Influences on performance

context in which he or she is placed. Probably none of these relations will attract dissent. Most people, I think, will agree that the learner's attitude to a topic, for instance, affects the attention given to instruction about it and consequently the amount of knowledge that is acquired. However, even if they are not contentious there are at least two reasons why the relations of the figure should be considered carefully. The first is that most research into learning of science, at least until quite recently, has been based on only a part of the figure and so has treated learning in so over-simplified a fashion that it has not been sufficiently illuminating and has had a limited effect on practice. Most studies through the 1950s, 1960s, and at least the first half of the 1970s concentrated on the direct relation between instruction and learning, and excluded consideration of the other influences that the figure includes. Typically, investigators would devise two or more teaching procedures, one of which might be designated the 'control' and which was intended to provide a baseline against which the 'experimental' treatment could be measured. Students would be allocated to one or other of these treatments, preferably randomly, and their performances on a subsequent test would be compared. The comparison usually involved a statistical test of whether the mean scores of the various groups were 'significantly different'. Significant in this sense means not whether the difference in scores matters educationally, but whether it is too great to be unlikely to have arisen by chance, through accidental allocation of better learners to one group. Statistical significance depends not only on how far apart the means are but also on the variation of scores within each group. The larger this variation, the less likely the difference between the means will be rated as 'significant'. Thus in these studies variation between individuals is regarded more as a nuisance than a matter of interest, and in statistical texts is often referred to as 'error', a pejorative term which hardly encourages people to look into its source.

Refinements in design of experiments were aimed at minimizing error, and so some attention came to be given to characteristics of the individual learners, such as their abilities. Measures of intelligence or of prior school performance would be taken, and used statistically as 'covariates' with the aim of explaining away some of the 'error' and so making the statistical test more powerful, that is less likely to rate a real difference between the treatment groups as 'not significant'. Reasons why learners differed in abilities, attitudes, or knowledge were not considered, and nor were the mechanisms whereby these differences came to affect learning – it was sufficient that they did.

My criticism of the form of investigation that prevailed for so long is not meant to reflect on the capacities or motives of the researchers. I can hardly afford it to do so, since I was one of them. Indeed it is only through considering the achievements and failures of that style of research that scholars have come to treat learning as a more complex phenomenon, and have come to carry out the more subtle and sensitive investigations which I cite in later chapters.

The second reason for considering carefully the relations within figure 2.1 is that the earlier theories are not, as a rule, so specific about them and so they have not been subjected to debate. Consequently, although I asserted earlier that the relations in the figure might not attract dissent, key concepts and some relations may be missing from it. The figure should therefore be taken as an initial proposal, to be argued over, modified and developed. That development is important because the more comprehensive a theory of learning, the better guide it will be to the practice of teaching.

The arrows in figure 2.1 lead towards performance, which is divided between learning and other acts. The division is not as sharp as it might appear, since doing almost anything involves some learning. Even routine acts provide feedback which leads to fine adjustments to skills, while more novel ones such as problem-solving involve considerable learning. Nevertheless, people do distinguish learning from other acts, and because it is the central concern of the book a chapter is devoted to the act of learning.

Earlier chapters are concerned with learning as an outcome: what sorts of things are learned, how they are stored and related in memory and what is meant by terms such as 'understanding' and 'concept'. I have given a lot of space to these things, because their names and others, like 'abilities', 'knowledge' and 'performance', can mean very different things to different people. Failure to define terms adequately leads to errors in communication. For example, in Bloom's otherwise admirable technique for promoting 'mastery', the criterion for mastery is defined loosely: 'We are convinced that the grade of A as an index of mastery of a subject can, under appropriate conditions, be achieved by up to 95% of the students in a class' (Bloom, 1968, p. 4). Bloom, and the investigators of his claim, never specified what is required for a grade of A. Even with the best intentions, so loose a criterion can lead to debasement of the technique: assessment could be based on trivial tests of recall, examinations of any level of intrinsic difficulty, or subjective judgements. Apprehension of that possibility may have hindered widespread acceptance of the results of research which concluded that Bloom's technique is effective, and militated against its greater adoption in practice. I have tried, then, to describe learning clearly.

One of the determinants of learning and other performance represented in figure 2.1 is *needs*. I found it convenient here to adopt Murray's theory of needs (1938). As well as the physiologically based needs of water, food, warmth and absence of pain, Murray recognized a set of what he called psychogenic needs, including needs for achievement, blame avoidance and affiliation. It is simple to incorporate these ideas into a theory of motivation to learn.

I have tried to say something useful about the roots of needs. Figure 2.1 shows them as consequences of knowledge, abilities, attitudes and physical state. The influence of physical state on physiological needs is obvious. If a person is hungry, there is a need for food, or if in pain, for relief. The other relations are less clear but can be illustrated by a few examples. As I describe later, one component of an attitude is a set of beliefs. A saint who believes that salvation is to be found through holy works and abnegation of bodily desires will perceive different needs from a hedonist who holds to the slogan 'live for today'. Or, in a school setting, an able child who has developed confidence through steady experience of success may have different needs from one who has suffered many failures and consequently has a different attitude to learning.

Figure 2.1 shows knowledge, abilities and attitudes closely bundled together. Because the human organism is an entity it is not surprising that reductionist attempts, like the present one, to sort out distinct concepts occasionally run into difficulties. Knowledge, abilities and attitudes do merge with each other. The opinions that are a component of an attitude look like facts, and sometimes facts are shown to have been erroneous opinions. Abilities, too, often depend on knowledge. It is, however, traditional to think of these three concepts as distinct, so I have continued to treat them separately while describing their relations as clearly as possible.

An arrow goes from the conglomerate of knowledge, abilities and attitudes to learning and other acts. A major emphasis of the book is on the influence that knowledge and abilities have on learning. Gagné (1965) and Ausubel (1963, 1968) stressed the crucial role of prior knowledge in new learning, and it has become so widely appreciated that it is hard to realize now how little attention it once received. Ausubel presented the role of prior knowledge as essentially positive, vital for meaningful acquisition of new facts. Gagné, too, represented prior knowledge positively, proposing that new algorithmic skills could not be learned unless specific prerequisites were possessed, and that if they were then the acquisition of the new skills would proceed smoothly. Recently, researchers have found that earlier learning can have a negative influence on the understanding of scientific principles, for instance that out-of-school experience leads to Aristotelian views of dynamics which persist despite later formal instruction on Newton's laws, or that non-scientific meanings of words like 'force' or 'animal' allow students to construct different meanings for statements from those intended by their teacher who used the words in a scientific sense.

The subtle, strong influence of knowledge on learning has to be described in detail.

It is almost tautological to say that ability influences learning. However, what should be of interest here, and indeed may be contentious, is the representation of abilities not as a singular uni-dimensional construct like intelligence, which is so stable that it is pointless to try to change it, nor as the more-or-less inevitable acquisition in the course of time of the capacity to perform a limited set of mental operations, as is emphasized in Piaget's theory, but as a set of cognitive strategies, specific skills which can be applied to wide ranges of tasks and which can be acquired or lost. My model presents cognitive strategies as the essence of abilities and so as determinants of learning and other acts.

A major difference between this model and those of most, possibly all, other theories is my inclusion of context and the individual's perception of context. It is surprising that the role of context was neglected formerly, because in applied disciplines such as medicine, law and engineering, as well as education, context is recognized as having a much bigger influence on outcomes than it does in purer sciences. It does not matter much at what time of day or in what size beaker zinc and hydrochloric acid are brought together; they will react in the same way, with the same products. But the room, the time of day and countless other factors can influence greatly what happens in human affairs. Because little has been written before about any other aspect of context than direct teaching, my discussion of the interaction of context with the characteristics of learning in determining performance is one of the most speculative parts of this book.

I could have incorporated teaching with context, since the acts of the teacher are part of the surroundings of learners, which they have to perceive just as they do any other feature of their environment. However, teaching is so central in education that I separated it in the model and discuss its relation to learning at greater length than other aspects of context.

The model stresses that context influences performance only through the individual's perception of it. This is a vital part of a constructivist theory. Because people placed in the same context may interpret it differently, they may act in very different ways. Their perceptions of the context are determined by their physical and mental states. An obvious illustration of the influence of physical state is a situation in which one person is asleep and another is awake. They are bound to have different perceptions of the situation! As a less extreme instance, it can be appreciated that people in different states of health may perceive contexts differently: an alert, fit person may consider a social occasion such as a party one in which interesting things are happening and in which it is fun to take part, while someone feeling below par may perceive it as an unpleasant ordeal from which to escape as soon as possible. The mental characteristics of knowledge, ability and attitudes similarly determine what we make of a context. We even use the word 'perceptive' to describe someone who appears able to judge wisely what is happening in a situation. Obviously knowledge plays a large part in this. If you know a lot about the circumstances in which you find yourself, you are likely to perceive the situation differently from how you would see it if it were strange to you. Similarly, attitude colours one's interpretation of an event; you can readily see that at football matches and other partisan sports in which officials make subjective rulings.

Perception of context is important in learning because it determines what the individual thinks is the purpose of the learning. If the general context of schooling is seen as authoritative and repressive, a different style of learning will develop from that fostered by a situation seen as liberal and helpful. In the terms of the model, different perceptions of context will encourage development of different cognitive strategies, and different patterns of learning will follow.

So far I have written about learning as an outcome of the various influences of needs, perception of context and knowledge, abilities and attitudes, but of course learning rebounds on these factors by producing new knowledge, abilities and attitudes which in turn affect perceptions and needs. This is represented in figure 2.1 by the one arrow that comes back from the right, from learning. Perhaps this influence was present already, represented by the arrow from experiences. Experiences, after all, must include events which incorporated learning. The representation in figure 2.1 may be clumsy, but it should not obscure the point that learning is a continuous, interactive process, in that what is learned now is affected by what was learned earlier, and will influence in turn what will be learned later. And because it is contrary to the assumption that lies behind most common practice in schools, I should also emphasize that the arrow from learning asserts that abilities are affected by learning and are not so stable that they are a lifelong constant of the person.

There remains one arrow to refer to. Genetic inheritance is shown as joining with experience and learning in determining abilities. This book, however, contains nothing about the role of genetic inheritance. The omission is deliberate, because my aim is to present a theory of utility, one that can be used to bring about achievement of goals, and, given present values, nothing on a mass scale can be done about genetic inheritance. Huxley (1932) fantasized about genetic control in *Brave New World*, but practical attempts seem to be restricted to the miscegenation laws of Nazi Germany and South Africa and the recent American notion of a sperm bank with deposits from Nobel Prize winners.

Apart from genetic inheritance, the relations between the concepts in figure 2.1 are the essence of the theory. Some are given more attention than others. First though, the meanings of most of the terms have to be described in detail, commencing with knowledge.

3

Elements of Memory

Aristotle regarded the brain as a radiator that served to cool the blood (4th Century BC/1956). We give it a greater role as the centre of thinking and memory, but hardly know more than the Ancient Greeks about how it works. Although we know that certain parts of it deal with specific operations, such as the occipital lobe with vision and the frontal lobe with planning ahead, the sites of individual bits of knowledge have not been found. It is probable that they do not exist, that the brain does not work like a digital computer with a bank of memory locations each capable of holding one piece of information. Memory may not be a static pigeon-hole system, but rather a dynamic process which can only be described by a very complicated model. Since such a model is beyond the reach of our present knowledge, I must present a static one which, although it may be far from an accurate representation of memory, I can defend on the grounds that it is helpful in describing learning and useful in deriving effective principles of instruction.

The first question to deal with – which forms the subject of this chapter – is 'What sorts of things do people know?'. The next chapter then turns to the ways these elements of knowledge can be associated with each other, while later, chapter 9 describes how they are acquired and inter-related.

Although the physiological basis of memory is obscure, it is clear that it involves large parts of the brain and even the muscular system. There are not different stores for different sorts of memory. Hence the identification of different types of memory which I make here is a distinction of convenience, made because the types have different conditions of learning. The seven types of memory element which I will use in describing the learning of science are: strings, propositions, images, episodes, intellectual skills, motor skills and cognitive strategies. Brief definitions are given with examples in table 3.1.

Table 3.1 Seven types of memory element

Element	*Brief definition*	*Example*
String	A sequence of words or symbols recalled as a whole in an invariate form	'To every action there is equal and opposite reaction'
Proposition	A description of a property of a concept or of the relation between concepts	The yeast plant is unicellular
Image	A mental representation of a sensation	The shape of a thistle funnel; the smell of chlorine
Episode	Memory of an event one took part in or witnessed	An accident in the laboratory; the setting up of a microscope
Intellectual skill	The capacity to perform a whole class of mental tasks	Balancing chemical equations
Motor skill	The capacity to perform a whole class of physical tasks	Pouring a liquid to a mark
Cognitive strategy	A general skill involved in controlling thinking	Perceiving alternative interpretations; determining goals; judging likelihood of success

Strings

Strings are units that are not readily paraphrased. They are usually verbal, but can be composed of other forms, such as musical notes. Examples of strings are memories of telephone numbers, lines of poetry, proverbs, the Gettysburg Address, the Lord's Prayer, addition and multiplication tables and the monologues of guides at public monuments. The telephone number, the poem, the address, are not themselves strings: the memory of them is the string.

Strings are learned as a whole, by repetition. They can be extended by learning more, which is what usually happens with long strings such as the

Gettysburg Address: the learner keeps repeating the address from 'Four-score and seven years ago . . .', adding a few words at each repetition until the whole is absorbed. When strings decay through lack of use, we usually find they are forgotten in the reverse of this process with the end being lost before the start. Although they can be forgotten, their heavy over-learning through repetition can make strings very stable elements in a person's memory.

Strings are units. Recall of the whole is triggered by its first bits. For example, the word 'awake' almost always triggers for me a stanza from Fitzgerald's translation of Omar Khayyam:

> Awake, for Morning in the Bowl of Night
> Has flung the Stone that puts the Stars to Flight:
> And Lo! the Hunter of the East has caught
> The Sultan's Turret in a Noose of Light.

Sometimes one would like to be quit of a string like that, but they are not easy to forget. When people are interrupted in the middle of recall of a string, and they lose the thread – an appropriate metaphor – they generally have to go back to the start. Another instance of this unitary nature of a string is the difficulty one has in starting a well-drilled string, such as the Lord's Prayer, in the middle. Almost invariably one finds one has to run through it from the start.

Most people who study science acquire lots of strings. Chemical formulae are an example. Given the cue 'H_2 . . .', most chemistry students and teachers find that the rest of the string, something like '. . . O' or '. . . SO_4', flashes into consciousness. Physics formulae, too, may be strings. The cue 'v^2', triggers '$v^2=u^2+2as$' for me. Laws and definitions may be learned as strings: 'To every action there is an equal and opposite reaction'. Mnemonics are strings, occasionally used in science, to serve as cues to recall of other information. I recall one to help with a row of the Periodic Table: '*He*llo *Li*ttle *Be*ryl *B*rown *C*hewing *N*uts *O*n *F*riday'. There is, however, less emphasis on the learning of strings now than in the times when the ability to trot out Latin and Greek tags was the mark of an educated man.

Since they are learned through rote repetition, strings can be totally meaningless to their possessors. Once, when standing outside a cage in the London Zoo, I was startled by a Cockney voice saying 'Help! Get me out of here'. I spun round: it was a mynah bird. Although what it said was sensible in context, I presume it did not understand what it was saying. The term 'parrot fashion' is appropriate for such meaningless strings. One can be less certain about the meaning a string has for a human. An example of a near-meaningless one is that of a friend of mine who taught his two-year-old grandson to say 'Wie geht es ihnen', the German greeting. The child did not know it was a greeting, he just knew it was something grandpa liked him to say.

Strings are not necessarily meaningless to their possessors. One might learn the Gettysburg Address as a string, while fully appreciating its meaning either at the time of learning or earlier or later. Actors learn their lines as strings and can deliver them without comprehending them, but for any quality of performance they must interpret them through facial and other bodily movements and through timing and expression in speech. Such interpretation comes only when the actors understand their lines. Often strings fall somewhere between absolute meaninglessness and high understanding. For many years state schools in Victoria began each week with a flag-raising and oath-taking ceremony. During the oath the students repeated lines which included '. . . and cheerfully obey my parents, teachers, and the laws'. The word 'cheerfully' usually came out as 'chiefly', which perhaps was a more accurate description of their behaviour. On the occasions when I asked children about this oath, few comprehended its meaning fully: they knew it was a promise, and that it was to do with good behaviour, but were vague about its purpose.

Another instance of part-meaningfulness comes from my experience in teaching physics. At the time, students were expected to remember verbatim many laws and definitions, and occasionally they would be asked in a lesson to repeat these strings. One of these was the definition of the ampere: 'The ampere is that constant current which, when maintained in two infinitely long parallel rectilinear wires of negligible cross-section, produces a force between them of $2\pi \times 10^{-7}$ newton per metre of length.' It is characteristic of strings and their longevity that when I came to write that definition for this book I was able to do so without looking it up, even though it was more than 20 years since I last used it. I had had twelfth grade students learn this definition, and one day when I had asked one to repeat it, it occurred to me to ask whether he knew what 'rectilinear' meant. He did not, so he could not comprehend the full meaning of the string. However, it was not totally arbitrary for him. He knew it was to do with electricity, he had a picture in his mind of two long parallel wires, and he knew it was a definition of the unit of current.

A final example is a string that I know, four lines from Alexander Pope's *Essay on Man*:

> Go, wondrous creature! mount where Science guides,
> Go, measure earth, weigh air, and state the tides;
> Instruct the planets in what orbs to run,
> Correct old Time, and regulate the sun;

I think I understand this pretty well, as an encouragement to mankind to unravel the secrets of the universe. It contains references to relatively recent (Pope published the Essay in 1733–4) discoveries by Newton on gravitation which led to measurement of the mass of the Earth and an explanation of the tides as well as Kepler's laws, and perhaps to debate about the reform of the calendar which was to take place in Britain a little later, in 1752, with 11 days being skipped to bring the country into accord with the European nations which had adopted the present Gregorian system from as early as 1582. The only misgiving I have about my understanding is that in another couplet from the same poem Pope makes the famous exhortation:

> Know then thyself, presume not God to scan,
> The proper study of mankind is man.

Although I find in this a justification for becoming a psychologist rather than a physicist, this appears to me to run contrary to my first string, so perhaps I have missed Pope's deeper meaning. Nevertheless, each of the strings is more to me than an arbitrary run of words.

Meaning to any degree is not then an essential characteristic of a string. They can be learned with understanding or without. The distinction of strings from the next type of element, propositions, lies not in their degree of meaningfulness but in their unvarying, unparaphraseable nature.

Propositions

Like strings, propositions are generally expressed in words. They are descriptions of the properties of concepts, or statements about relations between concepts. Examples are: the sun shines, birds fly, glass is transparent, the orang-utan is a primate, dogs chase cats, Milton wrote poems, acids neutralize bases.

The difference between propositions and strings is that the former can be paraphrased. A formula like $F=ma$ may be a string, but if the individual can paraphrase it to $a=F/m$ or any other equivalent form, it is a proposition for that person. The following statements are expressions of the same proposition:

The weather in winter is often bad.
Winter weather is often bad.
Frequently there is bad weather in winter.

A logician may argue that the expressions do not all mean the same thing, but that is not important. What matters is that the owner of the proposition sees them all as expressions of the same element of knowledge. There is one internal relation, which can be expressed in various forms. The proposition is the internal relation, the element of memory, not the external expression. This point may be obscured as we go further because any example of propositions that I give must be printed and therefore take the form of expressions, so I give it here what emphasis I can.

Propositions make up a large proportion of people's memories. In colloquial terms, propositions are the things people 'know'. They are knowledge, the sorts of things people tell each other; the sorts of things that make up the bulk of texts, encyclopedias and even newspapers. Many examples can be given from science: acids are sour; the crust of the earth is only a few kilometres thick; grass is a monocotyledon; fish have gills; uranium has 92 protons in its nucleus; metals are malleable; lunar eclipses are caused by the moon moving into Earth's shadow.

All the examples I have given so far are statements that most people would accept as correct. They are known as 'facts'. Other propositions are less generally accepted, less demonstrable, and are called 'beliefs' or 'opinions': for example, chemistry is fun; mathematics is a hard subject; women are less capable than men at mathematics and science; Australians are uncultured; the Earth is a flat disc supported by four elephants standing on the back of a giant turtle swimming in a limitless ocean.

There is no sharp distinction between facts and beliefs, for facts are just the generally accepted beliefs. This is true, even of science. These were once facts: the sun, moon, planets and stars are fixed to crystal spheres which revolve round the Earth; and, rather closer in time to us, the proton and electron are the fundamental particles of matter. As far as memory is concerned there is no difference between facts and opinions. Both are beliefs which are stored similarly, and may be learned similarly. The distinction is worth making though, because it will be useful when we come to the topic of attitudes. We need to recognize, however, that the distinction is not a true dichotomy, for the degree of social consensus about a proposition is a continuum rather than a divide, and may differ from one culture to another or between groups within a society. For instance, the belief that the Earth goes round the sun was accepted by an educated stratum long before it was common knowledge.

Degree of social consensus is one facet of propositions which is worth keeping in mind. Another is how arbitrary they are, that is whether they are social mores or physical facts, socially or physically determined; or, if you like, determined by man or by God. For instance, the belief that there are four seasons in a year is arbitrary: the people of Milingimbi in Arnhem Land identify six seasons. The belief that the sun rotates is not arbitrary: it can be demonstrated in several consistent ways. It needs to be appreciated that all beliefs are social, in the sense that our concepts are terms that we have invented to describe the universe. The sun and rotation are constructs that we have made in order to communicate to each other information about the things we perceive, but as they are so uncontentious a pair of constructs we feel that they are not arbitrary themselves, and that the rotation of the sun is something everyone could be brought to see.

Social consensus and arbitrariness are not totally independent properties of beliefs. Usually an arbitrary belief is one that we would class as an opinion, but that is not necessarily the case. The proposition about the four seasons is an instance; we tend to think of that as a fact rather than as an opinion, yet it is arbitrary. Figure 3.1 contains examples of propositions that differ in degrees of acceptance and arbitrariness.

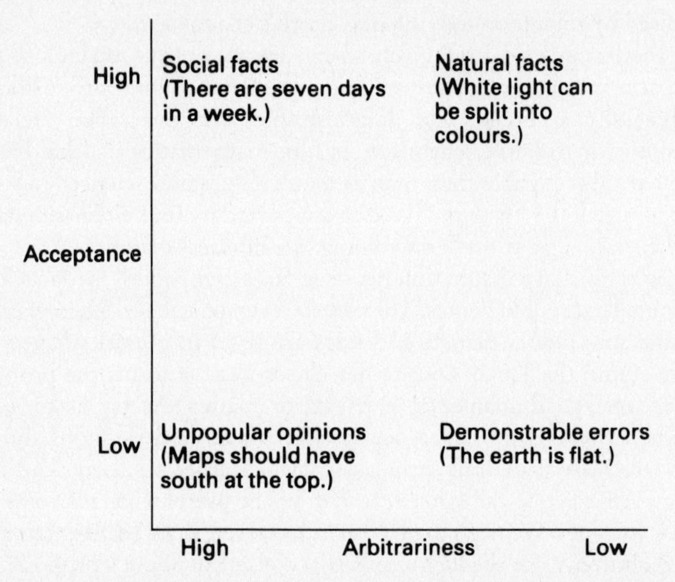

Figure 3.1 Variation of acceptance and arbitrariness of propositions

The propositions in the various corners of figure 3.1 tend to be of different types. The arbitrary, low agreement ones are often prescriptive, saying what *should* happen, or evaluative, saying whether something is good or bad. The ones in the opposite corner are more often descriptive. The separation is not perfect, however; figure 3.1 could be filled with more examples, making a continuous chain from one of these opposite corners to the other.

The properties of consensus and arbitrariness, or of prescription, evaluation and description, may not affect the storage of propositions, but are important when we consider how they might be acquired, and even more important in how they might be changed or discarded.

Images

The simplest way of describing images is to call them mental pictures. In figure 3.2 I have sketched images I have for several concepts. The mention of pictures is misleading, because it implies that images are solely visual, when they can be related to any of the five senses. If I am challenged to think of a red triangle, I can create in my mind a solid red one, or one with a thick red border, or 15 snooker balls at the start of a frame. In the same way, I can conjure up the feel on my fingertips of sandpaper or of silk cloth, or the sound of a bell, the smell of ammonia, the taste of sea water.

Figure 3.2 Visual images the author has for atom, flower and gum-leaf

Imagery was much discussed in the early days of psychology, from Fechner in the 1860s through Galton to Titchener early in this century. Francis Yates' fine history shows that as a notion it goes back to classical times (Yates, 1966). However, it has been a contentious topic in this century, and almost disappeared from academic view during the decades when behaviourism was dominant and introspection was condemned. Its renascence, in which Holt (1964) and Paivio (1971) were prominent, is one of the features in the recent rise in acceptance of cognitive psychology. Even in imagery's return, however, there has been controversy. Paivio (1969) implied that there are separate memory stores for verbal and imagic elements, which Pylyshyn (1973) disputed. Though of considerable importance in the development of theory of images, this issue is not as relevant to the function images play in the learning of science, and so I will not expand on it. I will merely postulate that images, however stored and however constructed, are a learned element of memory.

Images are often stereotypes, as my flower in figure 3.2 illustrates only too well. Although I am aware of some of the details of structure in flowers, and have seen many Australian flowers such as banksias, grevilleas, acacias and eucalypts, hardly any of which resemble the image in the figure, my stereotype is a relatively featureless approximation to some sort of standard European plant. That, I presume, is a result of social transmission from when I was a child. Many of our images are taught to us. One of particular importance in the learning of science is the image of scientists which people create. Cartoons transmit the notion of someone who is male, middle-aged to old, with glasses, eccentric, wearing a white coat and surrounded by strange objects. This image is likely to influence beliefs about scientists and science, beliefs that might be false and might be contrary to those we want people to acquire. I expand on this in chapter 7.

We are capable of inventing images for ourselves, as well as acquiring them from others. If I say 'Think of a chimney with a knot in it', you probably can create an image for it even though no one has shown you such a thing. I included the word 'probably' because people differ in their propensity and ability to form images. William James claimed he had no visual imagery, while Titchener clearly had a lot. It is difficult to believe that James really had none, because otherwise he would have had trouble recognizing objects and acquaintances. Results cited by McKellar (1972), obtained by himself, Marks and, 60 years earlier, Carey, indicate that nearly everyone reports some degree of visual imagery, with smaller proportions reporting the other sensory forms. It is clear, though, from the investigations of Marks (1973) and others, that people vary in the intensity with which they experience imagery. I first became aware of this in the course of a discussion with a colleague, who denied that people formed images at all. When I asked what she, as a physicist, saw when I said 'electron', she replied that she saw nothing. After much urging, she said she might see the quantum numbers for electron spin. Meanwhile I was seeing images of a spinning fuzzy ball, the atom pattern of figure 3.2, erratic dotted tracks as in a cloud chamber, and an *e* with a minus sign. Although I could create visual images much more easily than my colleague, I am equally limited with respect to olfactory images; I cannot think of the scent of roses, for example. People's imagery powers may vary across the senses.

The marked difference in visual imagery between my colleague and myself almost certainly means that we go about learning in different ways. Words in instruction, in conversation, in books, in thoughts, trigger many pictures for me of which she must remain unaware. In many circumstances I should be better off: I might find it easier to comprehend some topics, or having images for things might help me to remember information. However, great powers of imagery are not necessarily an advantage. Luria

(1965/1968) describes the fascinating case of a man who, through imagery, could recall arbitrary lists of words and numbers almost perfectly over intervals some decades long. This power was coupled with a debilitating interference with concentration on daily tasks and an inability to solve relatively simple problems – when the man started on the task or problem clouds of images would arise, stimulated by each new concept that came into his mind. Fortunately most people are between the extremes of my colleague and Luria's subject. For most of us images come to mind readily, but not to such an extent that they swamp other forms of thought.

Episodes

Cut in the stone above the entrance to the building by the botanical gardens in Oxford are the words that epitomize my feeling for the importance of episodes in the learning of science:

SINE EXPERIENTIA NIHIL SUFFICIENTER SCIRI POTEST

which I translate freely as 'Lacking experience, one can know nothing well'. Episodes are our records of experience, memories of events, occurrences we took part in or witnessed. The distinction which Tulving (1972) made between episodic and semantic memory is an important one to consider in the learning of science. Tulving emphasized that there is no separate store for episodic and semantic memories; the distinction is merely one of convenience for the analysis of memory.

Except for the small number of people with continuous amnesia, everyone has a large store of episodes. Our whole notion of self is bound up with our recollections of things we have done and experienced. Many of them date back to early childhood, though much of what happened to us then and later is irretrievably lost.

As an example of an episode, I can recall that when I was about 15 one of my cousins told me that it is possible to have acceleration without change in speed. This episode is worth considering for a moment, because it illustrates several points. The first is that though I recall it quite well, my cousin has no recollection of it whatsoever. Quite reasonably, he says that that is not surprising over so long an interval. But why do I remember it, and he not? Perhaps it was not an important event for him; he saw nothing unusual in it, nothing surprising, and so did not store it away. For me, though, it was a surprise: what he told me ran counter to my intuitive belief, and I did not want to accept it. When he explained how this apparent contradiction could be, I felt other emotions such as pleasure. There is also the possibility that he and I differ in propensity to store and recall episodes. I think we do.

Although this form of variation has not been studied systematically, almost certainly people do differ in this way, as they do in most others. The next point is that I have recalled this event several times in the many years since it happened and so have kept its memory fresh. It may be that only rehearsed episodes remain available for long periods. The nature of my recollection is also interesting. I recall it as a cameo; although I have a clear notion of where it occurred, and of details such as a book I was holding, I have no knowledge at all of what happened just before or just after the event. It is a few seconds sliced out of a time that is otherwise lost to me. The cameo comes back to me as an image, visual and haptic. I can see my cousin, feel the material of the train seats we were on, hear his voice. As well as these images, the cameo has verbal and emotive components. Thus episodes overlap the distinction I made earlier between propositions and images. They can incorporate both, yet are different because of the personal involvement, the belief that the event occurred. My final observation on this episode is that I do not really know that it did occur or whether I invented it and gradually came to believe in it, and if it did occur whether it happened just as I recall it.

The example I have just discussed concerned a unique event. We all have large numbers of such recollections; indeed they are often what people mean when they talk about their memories. The first time something of a particular sort happens to you, it is unique and you can store it as a specific episode. But what happens to that memory when much the same event happens again and again? Although you might be able to recall the original event, and any of the subsequent events which, though similar to the rest have some unique feature, the details of the mass blur, and you are left with a memory of the common features of the events. For example, breakfast: you might be able to recall one or two remarkable breakfasts, but the rest are not distinguishable although you can recall the general sort of thing that happens then. Schank and Abelson (1977) call these generalized episodes 'scripts', an apt term because they guide behaviour, helping people to cope with similar events when they recur.

Filtering a suspension is an example from science for which many people have a script. If you have done this often, it is possible to recall the script: folding the paper circle in two, then four, lifting up one layer and making a cone, placing the cone in the funnel, and so on. Scripts also apply to social behaviour. You know what to expect and what to be prepared to do when you go to the theatre, get on a bus, meet people, because you have scripts for these circumstances. We acquire many scripts. Without them the world would be a continuous puzzle, and every act would have to be thought out, or would have to be learned from someone step by step.

One reason for emphasizing the importance of episodes in learning is that teachers, because they are older, are more likely than their students to have scripts for situations, especially those that occur in school. Communication may fail when a teacher does not appreciate that students do not have a script for a situation, or misjudges the amount of time and help the students need to make sense of the situation and work out what to do. Another reason, to which we will return in chapter 8, is that people's perceptions of contexts are determined by the scripts they can apply to the situations.

Intellectual skills

Many theorists have overlooked the distinction that Ryle (1949) made between 'knowing that' and 'knowing how'. I have not, for instance, found any recognition of it in Ausubel's extensive description of learning (1968). Of course, Ausubel made it clear that his concern was 'meaningful verbal learning', which indicates a concentration on 'knowing that'. Theorists who have followed Ryle's distinction include Gagné (1968, 1972; Gagné and White, 1978), who used the terms 'verbal knowledge' and 'intellectual skills', and Greeno (1973), who used the terms 'propositional' and 'algorithmic knowledge'. 'Knowing that' is equivalent to my element, propositions; I have followed Gagné in using intellectual skill as the term for 'knowing how'.

Where propositions are unitary, each learned separately as a single fact, each intellectual skill is the capacity to perform a whole class of tasks. For example, once students learn how to solve Ohm's Law exercises, they can apply the procedure to any exercise of the type, even ones not seen before; or once they have learned to recognize liverworts, they can class new objects as liverworts or as not-liverworts.

Gagné (1965) described eight types of intellectual skill, but I choose to maintain only three of these. I prefer to see the highest skill in his system, problem-solving, as an act that involves application of memory elements, particularly cognitive strategies. Thus problem-solving is a complex performance rather than an element in memory. The lowest four skills in Gagné's system do not fit the definition I gave above, of the capacity to perform a whole class of tasks. His signal learning, for instance, consists of one-to-one correspondences, each having to be made separately in the same way as each proposition has to be learned separately. The three divisions of intellectual skills that I make are discriminations, classes and rules.

Discriminations

Discriminating is being able to tell which things are different and which are the same. A very simple example is the sort of task which used to appear in the *Sesame Street* television program: two identical shapes would be shown along with a third, different one (figure 3.3) and the question asked 'Which of these is not like the others?'. Although that task may seem trivial, it is a skill that has to be learned and is taken seriously in preschool programs. Toys such as boxes that have shaped holes through which similar shaped objects are to be fitted provide training in discriminating.

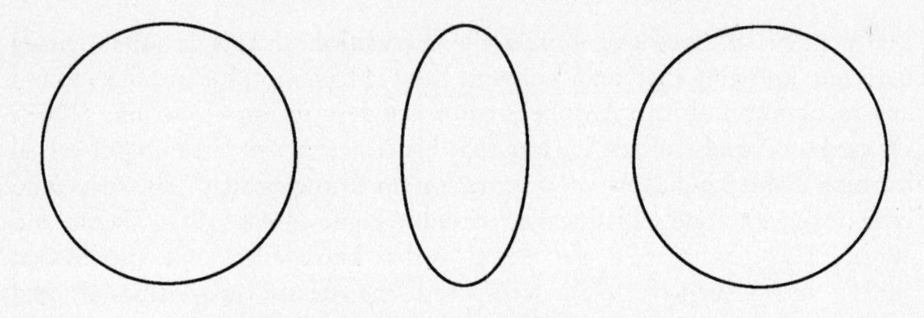

Which of these is not like the others?

Figure 3.3 A discrimination task

Discriminating may seem so simple that it is not thought necessary to consider it in higher levels of school or in the life of adults. It should not, however, be taken for granted. Subtle examples can be given of discriminations which, if unattained, may prevent further learning. A non-science example is the learning of tonal languages like Chinese or Thai. Thai speech involves the use of five tones, each involving a specific pattern of pitch. The rising tone, for instance, is akin to the expression of a question in English. The same basic sound may have a quite different meaning in each tone. Falling tone 'kow' is rice, rising tone 'kow' is white. Although I have lived in Thailand for several months in all, I am unable to discriminate these tones and so have found it impossible to make much progress with learning to understand or to speak Thai.

Another example, one that more people will have experienced, is being able to discriminate between wines. To the beginner they all may taste the same, while experts are able to make finer and finer discriminations.

It is not hard to think of examples from science. To the expert chemist not all white powders are identical, though at first they may be to students.

Two meter readings may look the same to a student, not to the teacher. To the students on an excursion, all hills may look much the same while the geology teacher sees them as being of several different types. Eucalypt fruits may not differ for the unlearned, yet are the basis of classification for the skilled. As long as all chemical equations look much the same, the student is not going to progress far in learning about them.

Discriminations can be a source of unexpected trouble in teaching. Once you have acquired a discrimination skill, it is difficult to comprehend that another person cannot see the difference that is obvious to you. It is easy for a teacher to overlook the problems students may have through lack of a discrimination skill.

Classes

Classing is a vital skill, possibly the element of memory we use more than any other. Everything we look at, hear, smell, taste, or touch, we try to class. We represent the world to ourselves in classes and in specific objects – those trees, my house. Even the specific objects are thought of as members of a class. Names and classes are attached to specific objects through propositions: 'That is mummy', 'Fred is a dog'. If we show a child an object and say 'That is a cloud', the child has a proposition that applies to that object. When the child generalizes the name to a class of objects, she has the skill of classing clouds. She can then class objects as yet unseen as clouds or not clouds when they appear. She does not need a definition of what a cloud is; that would be a proposition, not the skill itself. Sometimes the definition is a help, or even essential, to the execution of the skill, but it is not the skill itself.

Most of the classing we do is almost instantaneous on perception. We do not have to run through a list of properties each time to recognize objects as a tree, a car, a man, a road, a chair, a cup. We might even find it difficult to list the properties: how do you know that the animal you are looking at is a dog and not a cat? The classing is instant, the process probably involves judgements of general body shape, tail, head carriage and ears, but is essentially unconscious, and it would take some time to write out just how you recognize dogs.

Not all classing is by immediate perception. Sometimes we do apply a definition, comparing the object or act with a list of properties. We follow that procedure on two sorts of occasion: one is where although the object is capable of being classed by perception, it is of a class with which we are not yet thoroughly familiar; the other is where the class is never perceivable.

To illustrate the first case, consider a beginning student of chemistry who is asked whether a substance is a metal or a non-metal. The expert can tell at a glance, but the beginner has to run through a list of properties before deciding. As another example, when I am out in the country with a geologist I have to think before identifying (usually tentatively) features as horsts or faults and so on, or rocks as igneous, metamorphic or sedimentary, while my companion sees them immediately as members of these, and even more finely divided, classes. I check individual properties against a list, he sees the whole and classes it virtually unconsciously, as easily as I do for cats and dogs.

The other case is where the properties are not observable at all, or can be obtained only one at a time and indirectly. A class for which the crucial properties are unobservable is 'cousin'. You cannot tell by looking or through any other sense whether a person is someone else's cousin. Nor can you tell through direct use of your senses whether the gas in a jar is nitrogen or helium or argon. You have to apply tests to determine whether that substance is a pure metal or an alloy.

The distinction between these two cases is roughly the same as the one Gagné made between concrete and defined concepts (1977). Certainly the case of cousin is one of his defined concepts. The distinction should not be confused with one between concrete and abstract concepts. Justice is an abstract concept, but Socrates had no trouble in confounding people by asking them to define it. Yet we have no great difficulty in classing acts as just or unjust. We might differ in our classifications of instances, but each of us feels able to make the judgement. We do this all the time for abstract terms, readily classing events as kind, cruel, thoughtless, and so on. The distinction between concrete and abstract is not as useful in science as that between concrete and defined.

In science people learn many concrete classes: mosses, metals, magnets, forces, conglomerates, suspensions, dicotyledons, dispersion. They also learn classes which shade off into the unobservable or are distinguishable only by tests, such as electric fields, Group II metals, amphoteric oxides. In all cases the skill is one of recognizing whether the instance is a member of a named class. Once acquired it can be applied to new, previously unencountered instances.

As in other highly developed subjects, scientific classes can be divided into ever-finer classes. For example rock – igneous rock – granite – Harcourt granite; or living things – animals – vertebrates – mammals – primates – chimpanzees. Learning of these hierarchically arranged classes is not necessarily in one direction or the other. The first example above would usually be learned from the general class, rock, down to the specific, Harcourt granite, with many people learning granite before igneous rock. The second example is also likely to be learned erratically, with a common sequence being animal – living – chimp – mammal – vertebrate – primate. As learning proceeds, in whatever direction, finer and clearer distinctions are made between instances, and a more coherent classification system is acquired.

Classes are crucial in communication, for their names are the nouns and even the verbs and adjectives that appear in expressions of propositions. Thus if I try to tell you something about animals, you may receive an incorrect message if your class of animals is not the same as mine. We may differ because we really have idiosyncratic uses of the word 'animal', or because, although our uses are the same, we both have alternative uses for different contexts and you may be interpreting the message in another context from the one in which I framed it. Classes are a human fractionating of a continuous universe, so it is not surprising to find that reality causes trouble for our artificial system. Our use of language implies that boundaries of classes are sharp; we say this object is a chair, or is not, but we do not have an easy construction for things which are 'almost-chairs'. The fuzziness of boundaries allows different patterns of inclusion from one person to the next. For instance, would everyone include the following under the label of chair – armchair, stool, throne, sofa, bench, bean bag? You might want to argue that the object is a chair when it is used for sitting on, but does that mean that chairs exist only when being sat on, and the rest of the time are something else? And is everything that is ever sat on a chair?

Among scientists the fuzziness of boundaries and idiosyncratic use are not crucial issues, because there is a consensus. They are matters of concern in the learning of science, however, because learners are still acquiring the examples and information that bring them to the consensus. A teacher who has the scientists' consensus may be trying to transmit a proposition which is absorbed as something quite different by a student who uses the labels in the teacher's utterance to stand for other instances. For example, scientists agree on the use of the term 'force'. They will not disagree when shown some situation and asked about the forces present. The lay person's use of 'force' can be different. Osborne (1980) provides many interesting examples, one of which is shown in figure 3.4. When asked whether the policeman is putting a force on the demonstrator, one student said no, not if the demonstrator goes quietly.

Even scientists are affected by context in their use of words. After all, they are not only scientists but citizens, mothers, fathers, shoppers, too, and have to adjust their use of words to suit the situation. Frequently I drive on a freeway which has a large sign on it, 'Animals Prohibited'. I drive past it, and I have not yet been stopped for transgressing this by a policeman (another animal). I understand that animal in this context is a different usage from animal in a science context. The problem in learning is that students may not know which context is relevant. Errors in communication follow

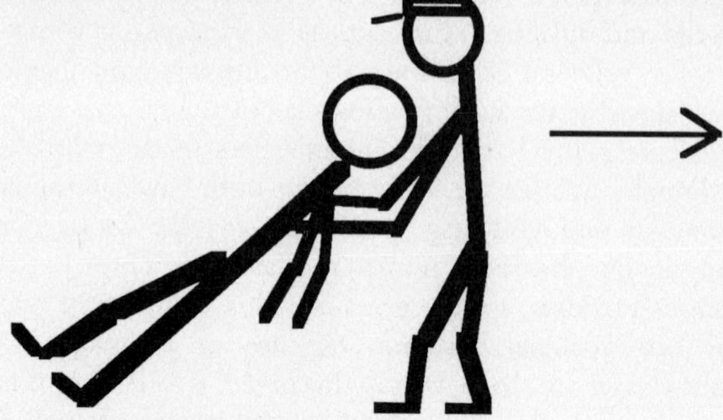

Figure 3.4 Line diagram used by Osborne (1980) in interviews on force

when the teacher uses a word in one context and the students interpret it in another.

Rules

Rule-following is the third type of intellectual skill. Rules form a part of our memories which is important in day-to-day matters as well as in the learning of science. Rules are procedures, algorithms, which are applicable to classes of tasks. Examples from science are being able to add two vectors, to find the velocity at a point on a position–time graph, to calculate the mass of one chemical needed to react completely with a given mass of another, or to balance chemical equations.

Rules are highly specific. They cannot be defined loosely, in a form such as 'solves momentum problems'. That definition is so broad that it will cover many types of problems, some of which a person may be able to solve without being able to do all. Rules must be defined so that, barring lucky guesses and slips through inattention, each individual consistently will be able, or unable, to perform any exercise which fits the definition. An example of a precise definition is:

> Given a pair of position–time axes, the intervals on which are marked in increments of 1 unit of length and 1 unit of time, bearing a curved graph, the learner calculates the velocity represented by the graph at a given point, with correct sign and unit.

This definition is readily translated to a task like that of figure 3.5. If the definition were changed, say to allowing non-unit increments on the axes, it would be a different skill, more learning would be necessary, and the test item would be different. Often, of course, when one skill has been learned there is little effort required to learn a closely parallel one, but there is some. An apparent inability to learn may really be an inability to transfer and to work out for one's self how to perform a skill. Research on learning hierarchies (e.g. Gagné, 1962; Trembath and White, 1979) shows that there is reason to expect that anyone with certain basic skills can learn any rule-following skill. It is a matter of analysing the rule into its constituents and ensuring that each is mastered in turn.

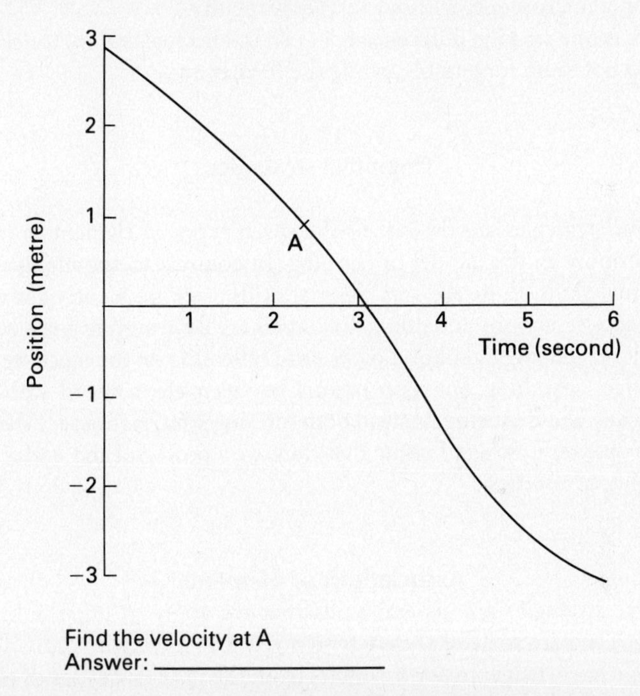

Find the velocity at A
Answer: ____________________

Figure 3.5 Test item for the intellectual skill of finding the velocity at a point on a position–time graph

Motor skills

Science could be learned as an abstraction, but it has always been a typically human activity in which the hand has had a part, as well as the brain and the eye. In the course of studying science people are likely to acquire some specialized physical skills such as making sections for microscope slides, using a pipette, ruling tangents to graphs. Also, the practice of science may promote other motor skills which are useful in non-science applications, such as the pouring of liquids to a mark.

Motor skills are memories of how to make muscle movements of a complex form. They share the attributes of intellectual skills of being analysable into prerequisite movements and, once learned, of being applicable to a class of tasks. An illustration of the application to a class of tasks is the skill of turning a knob. At an early age children learn this quite complicated movement of fingers and wrist, usually on door knobs or water taps. They can then apply it to bottle tops, screw terminals, or the focusing wheel on a microscope, without further difficulty.

There is one striking difference between intellectual and motor skills: the latter do not seem forgettable, while the former are.

Cognitive strategies

Cognitive strategies are the last of the seven types of element in memory that I propose in this model of learning. In contrast to the highly specific intellectual skills, these are very general skills, each frequently activated in diverse acts of learning and doing. Examples are determining goals, working out options, judging likelihood of success, reflecting on the meaning of new knowledge, searching out associations between elements of knowledge, generalizing and deducing. Instead of describing strategies here, I shall leave them to chapter 6, where I argue that they are a powerful and useful way of conceiving of abilities.

Associations of elements

Cognitive strategies are general skills that we apply in our thinking and learning. They are so general they tend not to be linked with individual bits of specific knowledge. In contrast, bits from the other six types of memory element can be linked together into conglomerations of knowledge that we call our concepts. The next chapter describes how those elements can be linked, and discusses how patterns of associations are related to the important but nebulous notion of understanding.

4

Associations of Memory Elements

Although your memory is a whole, with all of its parts connected, some bits of it are more nearly related than others so that we speak of your knowledge of this topic or that, meaning knots of memory elements that all refer to one subject and are closely joined to each other. But what is it that joins them? You could associate two bits of knowledge because they have often been presented together, just as Pavlov's dogs came to link the ringing of a bell with the giving of food, but that is hardly an important process. Nobody (one hopes) teaches science that way, for conditioned linking would be meaningless, contributing nothing to your understanding of either bit of knowledge. More likely, and more importantly, you link bits of knowledge meaningfully. The question is, what is the nature of those links?

For the six types of memory element, other than cognitive strategies, that were described in chapter 3, there are 21 combinations (including like pairs) of two elements to be considered in associations. It will not be necessary to describe all 21, as the forms of association will become clear from a few examples.

Consider first association between two propositions. The first pair in figure 4.1 is associated in memory if the individual is aware that they contain the common term 'base'. If one of these propositions is recalled, then the chance that the other will come to mind is higher than the chance for unrelated propositions, since the common term will act as a cue to its recall. Through associations between pairs of propositions a whole network can be formed, as figure 4.2 illustrates for the nine propositions of figure 4.1. In figure 4.2 a line links the numbers of two propositions when they share a term.

Strings can be linked with each other or with propositions if they can be dismembered into separate labels. I have a string 'In fourteen hundred and ninety-two, Columbus sailed the ocean blue'. Since I can pull the label

1 Acids neutralize bases.
2 Alkalis are soluble bases.
3 Acids contain hydrogen.
4 Active metals displace hydrogen from acids.
5 Strong acids are completely dissociated in water.
6 Most metal oxides are bases.
7 Alkalis feel soapy.
8 Acids turn litmus red.
9 Alkalis turn litmus blue.

Figure 4.1 Illustrative set of propositions

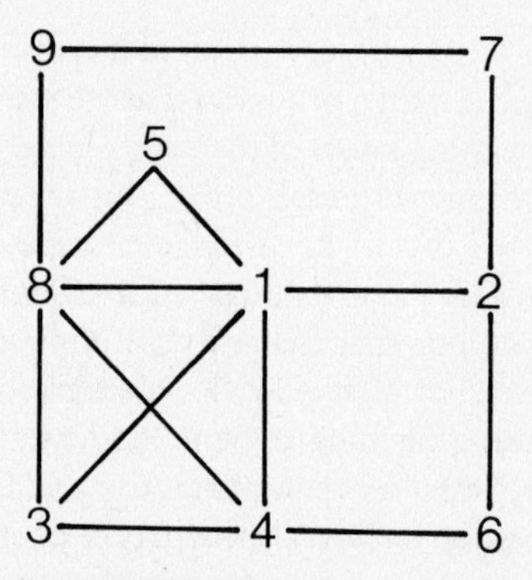

Figure 4.2 Pattern of links between the propositions of figure 4.1

'Columbus' out of this string, I can link it with any propositions I have with Columbus in them.

The other types of element need verbal labels if they are to be linked through common terms. I have an episode which can be put in words: somewhere near Titusville in Florida I saw a full-scale replica of Columbus's ship, the Santa Maria. Actually I do not recall this episode in words, more as an image, but there are words attached to it like Titusville, replica, Columbus and Santa Maria, and any of those words can trigger its recall. Some of the words are more effective than others. 'Replica' is less likely to promote recall of this episode because it is a part of many other episodes, propositions and images. 'Santa Maria' is very specific, and I can hardly hear or see those words without the episode coming to mind. The labels that are part of the episode are tags that enable me to link it with propositions and other elements.

I also have an image, derived from some now-forgotten source, of the head of a man with shoulder-length hair and a cap with the brim divided into four segments and turned up all round. This image is labelled 'Columbus' in my memory, so I can associate it with all the propositions that contain the term Columbus, and all the episodes that have Columbus as one of their labels. Images involving other senses can be labelled too, though we have fewer terms for experiences of smell and taste, in particular, than of sight. Non-visual images are often linked with, or are an integral part of, an episode – whether you call it linking or being part of depends on how you define episodes. These images can then be powerful cues to recall of the episode. A stray scent, one combination of sounds picked out from a background of noise, can bring back with startling effect an episode long unrecalled. Novelists put this better than psychologists:

Smell is of all senses by far the most evocative: perhaps because we have no vocabulary for it – nothing but a few poverty-stricken approximations to describe the whole vast complexity of odour – and therefore the scent, unnamed and unnameable, remains pure of association; it cannot be called up again and again, and blunted, by the use of a word; and so it strikes afresh every time, bringing with it all the circumstances of its first perception. This is particularly true when a considerable period of time has elapsed. The whiff, the gust, of which I speak brought me the Diana of the St Vincent ball, vividly alive, exactly as I knew her then, with none of the vulgarity or loss of looks I see today. (O'Brien, 1972, p. 240)

To take a less romantic example, the words 'hydrogen sulphide' may trigger an olfactory image. The words have been attached to the scent, normally through someone saying that the smell is caused by hydrogen sulphide. Once it has the label, the image of the scent can be triggered by recall of a proposition containing the label, or the smell may trigger recall of the proposition.

Intellectual skills can be linked with other elements if they have labels. They do not always have one, either because it has been lost or because the skill was learned without a label. I learned the skill of converting sums such as $7 \times 6 + 7 \times 14$ to 7×20 without knowing what it was called. I did not even give it a name of my own invention. Eventually I did find out what it was called, and so labelled it, but now I find I have forgotten the label so it is again unlinkable with other elements in my memory. Often, though, skills will have labels, which are not necessarily names. I can convert year numbers from the Christian system to the Buddhist. This is a simple skill, of adding 543 to 1988 or whatever the year is. I have labels on this, 'Christian year', Buddhist year', '543', any of which is liable to cue the skill for me and also enable me to link the skill with other elements of knowledge.

Much the same points apply to motor skills as to intellectual skills. Cues such as 'swimming', 'blowing on a carbon block', or 'making a transverse section' can stir muscle memories. The cues are labels for activities, and enable the motor skills to be associated with other elements. The association is, as in other cases, two-way. A proposition containing the label may cause recall of the motor movement; or going through a movement, exercising the skill, may cause recall of a specific episode or any other sort of element.

Cognitive strategies are less part of the pattern of associations, because for most people they have no label; they are simply part of their thinking. Even if they are named, they are so pervasive a part of mental functioning that they tend to be not specifically connected with any one piece of knowledge. There are exceptions, however. If at some time you apply a strategy in some remarkable way, or observe someone else doing so, the strategy can be linked to the episode. Then whenever you recognize that you are applying the strategy, the episode is likely to be recalled.

In sum, elements of memory are connected through labels. Propositions and strings may contain their labels as well as having others attached, while other elements will need to have labels given to them.

Labels can bind sets of elements into a network. For me the label 'Ohm's Law' is attached to a string, the statement of the law which I learned by heart as a schoolboy. That label connects the string to elements that share it: a proposition that $V=IR$ or $R=V/I$, an image of V above IR inside a triangle, and the skill of being able to find any one of V, I and R when given the other two. Many other propositions and other elements are connected to these elements in turn. The propositions of figure 4.1 are associated for me with images of bottles with glass stoppers, the sharp smell of hydrochloric acid, the greasy look of concentrated sulphuric acid, the lively taste of carbonic acid. I also recall generalized episodes of titrations using phenolphthalein as an indicator, and the motor skill of pouring concentrated acids from large Winchester flasks into other containers. I would not be able to join these images and episodes with the propositions if I had not attached the label 'acid' to them. Of course each image or episode may have several labels, like 'burette', 'titration', or 'indicator'.

As one further example of association, let us imagine the sort of knowledge that might be lodged in the memory of a middle-secondary school student soon after a series of lessons on density. A prominent part will consist of propositions and strings, which can be listed like this:

1 Density is the mass per unit volume of a substance.
2 $D = M/V$.
3 Lead is very dense.
4 Polystyrene foam has low density.

5 The density of water is 1 g cm^{-3}.
6 Ice is less dense than water.
7 Solids float in liquids that are denser than they are.
8 Icebergs float.
9 Mercury is denser than lead.
10 Lead floats in mercury.
11 Gases have low density.
12 Hydrogen is a gas.
13 Hydrogen has the lowest density of all.
14 Bubbles of hydrogen float in air.
15 Oil floats on water.
16 Oil is less dense than water.

Most of these propositions are linked to each other by the term 'density', which they contain, but there are other terms that link a few elements each. Numbers, 3, 9 and 10 contain the term 'lead', and numbers 7, 8, 10, 14 and 15 contain 'float', so these elements form closely associated knots.

As well as the verbal knowledge of propositions and strings, the student may have acquired a couple of intellectual skills:

17 Given values of two of mass, volume and density, can calculate the value of the third.
18 Given two objects, decides after hefting them which is denser.

The first of these skills is no more than substituting in the $D = M/V$ formula. It is likely to have been learned through direct teaching and drilling in its use. The second could also have been taught directly, but is more likely to have been acquired through experience in hefting objects. The student's teacher could have planned for the formation of useful episodes by giving the student unusual objects such as a (well-sealed) bottle of mercury to handle. Because of previous experience with bottles of about that size the student would have been surprised by its mass. Teachers can often use surprise to promote useful episodes. I used to have a litre milk carton filled with lead and leave it standing on the bench. The casual instruction to a student to get it and pass it round the room led to surprising physical experiences for the class, and, as I intended, to the students' constructing long-lasting episodes linked to density. At the other end of the density scale I had a large block of polystyrene foam painted silver-grey so that it looked like metal. I used to stagger in with this and pretend to trip, and would drop it on a student. The shock and then the unexpected lightness was intended to promote an unforgettable episode. (Fortunately none of the students on whom the block fell had a weak heart. I would be more cautious now, in this more litigious age.)

As a consequence of planned and unplanned experiences our hypothetical student might have these episodes related to density:

19 I remember picking up a lump of polystyrene and being surprised by its lightness.
20 I remember the teacher making bubbles with hydrogen gas soap solution. It was outside and the bubbles went up a long way before they burst.
21 I remember grabbing at the milk carton full of lead and nearly straining my arm.

The student's memory may also contain images, such as (22) an iceberg and (23) a picture of dots representing atoms, close together for a dense substance and far apart for a less dense one, and perhaps (24) a haptic one of the feel of a dense object as it is hefted.

We can imagine that some of these 24 elements were taught to the student, either by direct telling or the arranging of an experience, while others were learned more casually. In whatever way they were learned, they constitute the student's knowledge of density. This brings us to a term that is widely used in education, the notion of *concept*.

What does it mean to say that someone has the 'concept of density'? Can we say that this person has the concept and that person does not? In my model concept is the collection of memory elements that are associated with the label (density in the example), and the pattern of their links. Conceiving of concept in this way allows for two people to have any degree of similarity or difference between the meanings they have for a concept. Their concepts are the same if they have identical sets of images, propositions, episodes and so forth, about the label, but usually they will not, even if they have just been through the same course of study. The similarity of their concepts is a function of the overlap, the intersection, between their sets of elements.

There is, then, no simple answer to the question 'Does X have the concept of density?', unless X knows absolutely nothing that could come under the label of density, when the answer is a simple no. There cannot, however, be a simple yes. Whatever X knows about density is X's concept of density. The question really means 'Is there sufficient intersection between X's concept of density and mine (or what I think people like X should know) for me to judge that X knows enough about density?'. This argument implies that it is subjective whether someone can be said to have a concept or not. It might be asserted that a definition is an essential part of a concept, but even that can be disputed, as I tried to show in chapter 3 in the instance of recognizing cats and dogs. It should also be clear that possession of a concept is not a dichotomy in the sense that either one has it or one has not. Since the concept is the set of related elements, and these can be added to virtually without limit through new propositions and new episodes, it is really a zero-infinity rather than a zero-one situation. It is the *elements* that are possessed or not possessed; the concept is possessed to a greater or lesser degree. A third note is that two people can have extensive knowledge of a concept without having many memory elements in common.

This issue of what is meant by a concept has an important practical application. Customarily, courses of study are set out in terms of labels of concepts. This is done because it is convenient, much less tedious than listing or even thinking about the many propositions and skills that the students should acquire. It is a practice that causes no great problems when the constructor of the syllabus is the teacher. The concept label is simply a shorthand representation of all the elements that the teacher has in mind now, or believes he or she will be able to identify later, to transmit to the students. It does cause trouble, however, when the constructor and teacher are not the same person, such as when the constructor is a coordinator or head of department in a school and the teacher is an assistant; or on an even more remote scale, when the constructor is outside the school altogether, such as a syllabus committee responsible for an external examination or for a recommended course. In that case a consensus gradually develops over the range of propositions and skills that the term covers. This consensus is carried in the heads of the people involved and not written because of the labour as well as disputation about detail that could be involved. New members of the system are therefore at a disadvantage until they have built up their own interpretations of the terms through experience.

Possibly an even more important consideration is that teachers and students will have different meanings for a term, which can impede accurate communication between them. Generally the teacher will know more, but only rarely will the student have no knowledge about a concept other than that shared by the teacher. Students are almost bound to have their own episodes, and also may have esoteric beliefs, some of which are incorrect. Bell's study of people's notions of animals (1981) provides a basis for speculative illustration of this. She found that everyone in the wide spread of age groups she interviewed classed cow as an animal, but fewer classed man, whale, worm and spider as animal. We might guess that many of these people had propositions like 'animals have four legs' and 'animals are furry', which would not be shared by a biologist. If the discrepancy between the teacher's and a student's meanings is large, then communication is likely to be misunderstood in either direction, particularly when the student has much knowledge that is not shared by the teacher. In that case, the student's interpretation of what the teacher said is likely to be quite different from what the teacher meant. In any communication there are two acts of construction: the giver constructs words in sentences to represent the meaning of what he or she knows; the receiver interprets those words to store the meaning he or she places on them. Where the words mean different things to giver and receiver, the message must be different from what was intended. Where the teacher does not share the student's meanings for words, she cannot allow for it, cannot phrase the communication so that there is no possibility of misunderstanding it. Knowledge that the teacher has but the student does not is less of a problem, provided the teacher is aware of it and frames messages in a form that the student can understand.

The foregoing discussion touches on two functions of teaching. The easier function is to reduce the amount of knowledge the teacher has but the students lack. The teacher shares propositions, skills and images with the students so that their knowledge grows. Many simple models of learning see that as the only function of teaching. A more difficult, and more overlooked, function is to remove from the knowledge that the student has but the teacher lacks those propositions that are erroneous and are likely to inhibit subsequent learning and the acquisition of proper control and understanding of the learner's environment. These two functions may well require different teaching procedures. Before coming to those procedures in the final chapter, however, we should consider another property of knowledge, that of understanding, and then some characteristics of people which determine the quality of their learning.

5

Understanding

We can teach our students a lot of science, and drill them in this knowledge so that they can use it in passing tests, but do they understand it? Certainly we want them to. There is general acceptance, without debate, that students should not only learn things but should also understand them. Understanding is one of our more widely used words, both in normal language and in education. But although the word is widely used, and understanding is so valued, it is a term that is not well defined. Indeed it is not easy to define, though any psychology of learning should grapple with this issue.

Understanding can be defined as the ability to use knowledge, to cope with situations. That definition is behind the use of problems in school tests, and of transfer tasks in research, as measures of understanding. But really the definition and the tests are dealing with an overt performance, not an internal state. The thrust of my model is to go behind the overt performance and describe the organization of knowledge that supports it, defining understanding in terms of elements of memory and their pattern of association.

Part of the difficulty in defining understanding is that it takes on different meanings, depending on the scale and nature of what is to be understood. We can talk about whole disciplines: 'Does he understand physics?'. Or about concepts: 'Does she understand force?'. Or about single elements: 'Does she understand what is meant by force is a vector?'. Or longer communications: 'Do you understand what this chapter is about?'. We can also talk about situations: 'Does he understand this problem?', and 'Does he understand what is happening here?'. Or about people: 'I don't understand my children'. The last of these, understanding of people, is not part of the psychology of learning of science so I will say little further about it. The others need to be considered in some detail.

Understanding of concepts

The previous chapter ended with a discussion of the meaning of concepts, and therefore it is convenient to consider their understanding first.

Language influences the way we think about understanding: the question 'Does he understand density?' has a form that asks for the answer to be yes or no, when, as I argued in the previous chapter for the question about whether a person has a concept, it is really a matter of degree. I stated there that the person's concept of a notion such as density is that person's collection of propositions, skills and other elements that are linked closely to the label 'density', and pointed out that the collection can grow without limit. Just as the answer to the question 'Does he have the concept of density?' could be answered only by 'to some extent', so must the same answer be returned to the question about understanding. It is a subjective question – the answer depends on the questioner. The questioner has a concept of density, that is, his or her own collection of elements. The questioner wants to know whether the respondent's concept contains the subset of elements which the questioner will accept as reasonable for someone who can be graded as 'understanding'. Usually the questioner will possess this subset, though occasionally the question is asked in a genuine spirit of seeking information, from someone who is accepted as an expert, about a third person. For example, 'Does the President understand the physical effects of radiation?'. More often, though, it is asked in order to make a judgement. What one questioner has in mind as reasonable may not be the same as another's requirements. Further subjectivity comes when the questioner adjusts the requirement for different respondents. In making my judgement of whether someone understands 'density', I would have a different set of elements in mind for a science graduate than I would have for a 13-year-old.

No doubt most science graduates know more about density than most 13-year-olds, but here language again misleads us. Degree of understanding sounds linear, but it is not a simple matter to say that one person has a greater understanding of a concept than does another. Usually the relation between their sets of elements will be overlapping, as in case (a) in figure 5.1, and not inclusive, as in case (b). For case (a), it is a subjective judgement whether the elements $\bar{A} \cap B$ are more valuable in understanding than those in $A \cap \bar{B}$. Even in the apparently more clearcut case (b), where A knows all that B does and more, it depends what is in $A \cap \bar{B}$, the area outside the common knowledge. The elements there may be ones we assess negatively, beliefs that we judge reveal lack of understanding rather than additions to it. If A believes 'Density depends on shape', we might regard this as making A's understanding of density less satisfactory.

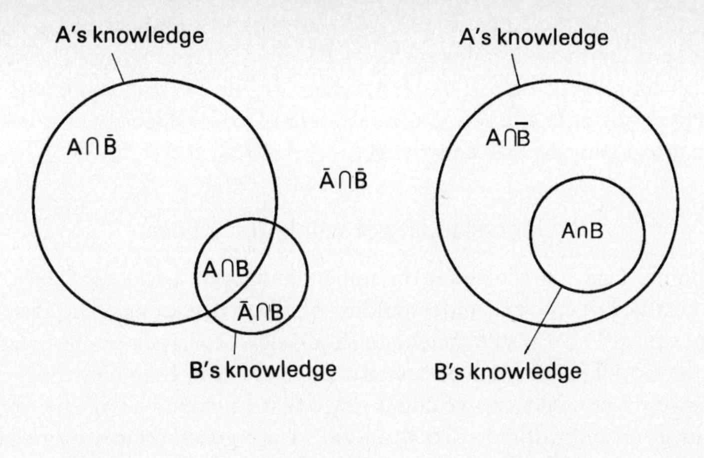

Figure 5.1 Hypothetical relations between the elements of knowledge possessed by two people

It is not just the amount of knowledge that matters, though naturally greater knowledge tends to engender greater understanding. The nature of the knowledge and the pattern of associations between its elements are important too.

The relative proportions of strings, propositions, skills, images and episodes affect the quality of understanding. A person whose knowledge of a concept is almost wholly propositional has a different form of understanding than someone with many images or episodes. Episodes may be particularly important in understanding, as they give a feeling of confidence in the accuracy or credibility of the knowledge. It is one thing to learn the proposition that metals expand when heated, another to see it happen, especially on a large scale as in the gaps in railway lines or in the expansion plates of bridges. This is put succinctly by Gordon (1976):

> Once one has watched the Brownian movement one's apprehension of the *nature* of heat will never be the same again. It is not that one can be said to have learnt anything in an objective scientific way but rather that one has come to terms with the kinetic theory of heat at a subjective level. It is the difference between having a sunset described and seeing one. (p. 223)

The importance of episodes in understanding is the theoretical justification for the emphasis placed on demonstrations, laboratory work and field trips in science teaching.

'Pattern of association' refers to the extent of links between elements. One's knowledge of a topic may consist of many sparsely connected elements, while the same number of elements in another person's mind could be bound into a coherent mass by many shared terms. The second pattern represents better understanding. In short, understanding of a concept is not only a function of the extent of knowledge about it but also of the integration of that knowledge.

Understanding of whole disciplines

The points I have made about the understanding of a concept apply, with even greater force, to the understanding of whole disciplines. The answer to the question 'Did Einstein understand physics?' can only be 'to a degree'. The question 'Did Einstein understand physics better than Newton?' might be answered yes, because we could judge that Einstein had all the relevant propositions and skills that Newton did, plus a good number more gained from his own work and the discoveries others had made in the intervening centuries. Einstein's and Newton's images and episodes would have differed, but we might accept that this would not constitute a vital difference in their understandings. The question 'Did Einstein understand physics better than Bohr?' is much less easy to answer. We might value differently the contributions of each to knowledge, but it is hard to say whether the things that Einstein believed about physics and which Bohr did not are more central to understanding to the subject than those that Bohr did not share with Einstein.

The view I have put forward implies that there is no central core of knowledge which is essential to the understanding of a discipline or even of a concept. That is not to say that all knowledge is of equal value or relevance in understanding. The judgement of that relevance must, however, be subjective. I am prepared to say that the proposition 'A force is a push or a pull' is more central to the understanding of force than 'There are four fundamental sorts of force in the universe', but others may not agree. In the same way my view that knowledge about velocity and acceleration is more important to the understanding of physics than knowledge about quarks might be challenged. Incidentally, Auguste Comte's opinion was that 'A science cannot be completely understood without a knowledge of how it arose' (1855, p. 43), a view that is not given much weight in many current science courses.

In sum, understanding of a concept or of a discipline is a continuous function of the person's knowledge, is not a dichotomy and is not linear in extent. To say whether someone understands is a subjective judgement which varies with the judge and with the status of the person who is being judged. Knowledge varies in its relevance to understanding, but this relevance is also a subjective judgement.

Understanding of single elements

Some different points arise when we consider the understanding of a single element, such as 'Force is a vector'. This proposition contributes to the understanding of the concepts of force and vector, but cannot be understood, that is take on meaning for its believers or recipients, unless they have sufficient other elements associated with the labels force and vector for them to be understood to a reasonable degree. Each recipient must also have some meaning for the relation 'is a', but that is acquired so early in life by virtually everyone (in their own initial language) that it can be accepted as understood. Just as it was not possible to specify the essential elements for understanding of a concept or a discipline, it is not possible to say which elements must be part of the constituent concepts in a proposition for it to be understood. It is again subjective. I might hazard some elements for force and vector, such as those in figure 5.2, but there is no requirement for others to agree with me and indeed I am not at all convinced myself about the inclusions and exclusions involved in forming that list.

Understanding of another form of element, the rule-following class of intellectual skills, requires something more. Rules can be learned as a procedure to be followed without understanding, as is often demonstrated in mathematics classes. The procedure is less rote and more meaningful when labels are known for the concepts and operations in it, and those labels are connected to many elements. For example, the rule of finding velocity at a point on a position–time graph (see chapter 3) involves the concepts velocity, displacement, time interval, slope or gradient and tangent. The rule can be performed without knowing what these words mean, almost without seeing or hearing them, but is more comprehensible when they are understood. More central to understanding of the rule, however, is being able to explain the procedure. If the person knows why this is done, or that, then there is greater understanding. 'Why do you draw a tangent to the point where the velocity is to be found?' 'Because the tangent is the line which has the slope of the curve at that point (demonstrates with diagram), and velocity is the time rate of change of position, which is the slope of the graph.' These explanations must themselves be understood as propositions. They cannot be rote strings, learned as a catechism.

Understanding of motor skills is similar to intellectual ones. Explanation is the key. 'Why do you have to hold your left arm straight when hitting the golf ball?' 'Because otherwise the left hand has to move in an arc of smaller radius. That makes the clubhead cut inwards across the ball, giving it a sideways spin which bites on the air and makes the ball slice to the right.'

Propositions

A force is push or a pull
Force causes acceleration
Force is proportional to the time rate of change of momentum
$F = ma$
Force can act through contact or at a distance
Electrostatic, magnetic and gravitational forces act at a distance
Distant forces follow the inverse square law
Tension is the name for forces exerted by ropes
To every force there is an equal and opposite force
Reaction is the name for forces opposing imposed forces
The forces on an object can be summed up to one resultant force
A force can have a twisting effect about a point, called its moment
Vectors have magnitude and direction
Vectors contrast with scalars, which have magnitude only
Vectors can be resolved into components
When you add or subtract vectors you have to take directions into account
Vectors are represented by an arrow, its length represents the size of the vector and its direction the direction of the vector

Episodes

Pushing and pulling objects
Pulling on springs and elastic
Feeling magnets attract and repel
Using two or more forces at angles to move an object

Images

Arrows for vectors

Intellectual skills

Given two of F, m, and a can calculate the third
Given a diagram of a situation, can mark in the forces acting
When shown the forces on an object, can judge the direction it is accelerating
Can add two vectors
Can resolve a vector into two components

Figure 5.2 Elements of knowledge useful for understanding the concepts of force and vector

Note though, that as with intellectual skills, routine performance of a motor skill is possible without explanation or any depth of understanding at all.

Explanation and understanding

Explanation may be the key to understanding of intellectual and motor skills, and of propositions too, yet the nature of explanation itself is not easy to describe. I could say to explain is to give reasons for, but then I have to describe what is involved in giving reasons. In the end, explanation often comes down to putting things in more familiar terms, and not to pushing them to fundamentals. For instance, suppose you have the skill of being able to find the speed with which something hits the ground when it falls from a given height. I ask, how do you do it, and you reply, $v = \sqrt{19.6s}$. Why that? Because the definition of acceleration is rate of change of velocity, hence $v = at$, and distance is average speed $(\frac{1}{2}v)$ times time, so $s = \frac{1}{2}vt$, or $\frac{1}{2}v(\frac{v}{a}) = \frac{1}{2}v^2/a$, hence $v = \sqrt{2as}$, and $a = 9.80\ \mathrm{ms}^{-2}$, so $v = \sqrt{19.6s}$. I might be satisfied with that, a derivation of the formula you were using. But I might go on to ask why *a* is 9.80, and why do things accelerate when they fall, and why do they fall anyway? Eventually we would reach the reason, because of gravity. When I ask, what is gravity?, we have reached an end. You will not be able to explain what gravity is in terms other than those such as force, that have been used already.

The familiar terms which satisfy us in explanations are often based on direct experience or on concrete demonstrations. How can aeroplanes stay in the air? An explanation can be framed around Bernoulli's Law, but that is less convincing than a demonstration of lift by blowing on the edge of a light piece of shaped wood or a sheet of paper. Often we are satisfied with demonstrations, real or in thought, through models. Kelvin believed that nothing had been explained until it was demonstrated with a physical model. He had his assistants construct complicated models of atoms containing many springs which could be set vibrating to demonstrate what might be happening inside atoms. Classic models in the history of science are representations of light as waves or particles, which are concrete, macroscopic ways of demonstrating reflection, refraction and other properties of light. They do not really explain why light behaves as it does, they merely show how some of the phenemona of light can be simulated. We are happy with these demonstrations because we can see the processes occurring with familiar objects.

Understanding of extensive communications

So far I have discussed understanding as if it were a static property, depending on the extent, nature, and pattern of interlinking of knowledge. However, part of the difficulty of defining it is that the word is also used to refer to a state of mind, a feeling of mastery, and to a process, the act of comprehending. All three usages apply to understanding of concepts, disciplines and single elements, as they do to all the targets I am discussing. The emphasis, though, that we place on the usages may vary from target to target. When we talk about understanding of an extended communication like a conversation, a lecture, a chapter of text, a poem, or a painting, we tend to mean, rather more often than we do for the earlier targets, the act of comprehension, the process of making sense of what we are hearing or seeing.

The process of comprehension inevitably depends on understanding of the constituent memory elements that are formed or recalled on receiving the message. When a person reads a text, the message cannot be comprehended unless all, or certainly the great majority, of its sentences can be converted into propositions that the person is familiar with and understands, in the sense described in the previous section, or that can be acquired and related immediately to elements already present in memory. Imagery will often be part of the process of understanding. I can illustrate this by imagining the reactions someone might have when reading a passage from the PSSC Physics text (Haber-Schaim et al., 1971, p. 240) (see figure 5.3).

Comprehension will often involve more than splitting the communication into separate elements and checking that each is understood. Often the communication has to be classified. Is it a direct statement or an allegory? Is this transmission of new information or elaboration of old? Are the values of the communicator relatively neutral or highly biased? It may also involve determining structure, seeing how one part of the communication relates to another. Texts are usually organized so that classification and determination of structure is as straightforward as possible, so that comprehension largely depends on the understanding of the individual elements. This is not the case for all other forms of communication. A poem such as Blake's *Tiger* has depths of meaning that involve much more than decoding of its sentences, which are themselves quite simple. Many school students will read the poem at that shallow level, of course.

Passage	*Reaction*
Two parallel metal plates	(Image of the plates side on) Metal, yes, probably brass. Plates – flat circles, not like dinner plates.
are connected to a variable voltage supply,	A power pack (image of the power pack that is common in the lab, a grey metal box with a knob on the front for changing the voltage output). Presumably a wire goes from one plate to one terminal and another wire from the other plate to the other terminal (image of this).
and a sensitive current meter	A galvanometer (image of galvanometer and of the symbol used for it in circuit diagrams).
is inserted in one of the connecting wires.	Funny way to put it. Sounds like the galvanometer is poked up a hollow wire, when they mean it is joined to the plate and to the power pack by two separate wires.
The plates are at a distance of several centimeters,	From what? Each other, it must be.
and the space between is filled with the gas under investigation	There's gas between the plates. How do they do that? The plates must be in a box of some sort, metal, pump the air out and flood the new gas in (various images).
When ionizing radiation (X-rays or ray from a radioactive source) passes through the gas,	Plural, singular. OK, I see, it is the radiation that passes. They shoot X-rays or something across the space between the plates (image).
the meter shows a current . . .	The galvanometer needle shifts (image).
As the potential difference between the plates is increased,	Turn the knob.
the current increases rapidly at first, then more slowly, and eventually reaches a certain limiting value called the saturation current.	(Image of galvanometer needle swinging quickly to a high value, then creeping on a bit more. Image of a graph of the current versus some quantity on the horizontal axis which could be potential difference or time).

Figure 5.3 Example of decoding of a passage of text

Understanding of situations

By 'situations' I mean the complex set of events with which one is in contact at any time. Much the same points that were made about comprehension of a communication apply to situations, which can be thought of as instantaneous communications in which all the information is presented instantly rather than sequentially. I do not mean to imply by this that situations cannot be dynamic, a succession of scenes that unfold one after another; rather, I am pointing out that situations are often less focused and require a gestalt appreciation of the whole pattern of stimuli and also selecting and discarding of the stimuli, processes which are only slightly involved, if at all, in a communication.

The process of comprehending the situation shown in figure 5.4 could go as follows:

> A man, holding a block of wood. The wood has a string on it, going over the wheel – it's a bike wheel – to a bucket. Perhaps there is something heavy in the bucket. The block looks heavy. Presumably the wheel can turn. It's a pulley. There is no tyre on the wheel, so there'll be a groove there for the string to fit in. The string looks taut. Is that because the man is pulling down on the block? If he is, the bucket must weigh more than the block. Could he be holding the block up? Yes, if the block is heavier than the bucket, he will be. I can't tell, though he doesn't look like he is lifting it. If he let go, I would see. The block and the bucket would then accelerate in one direction or other until one hits the floor. This is an odd picture. Why is he doing this? It must be a physics demonstration of some kind.

This example illustrates how comprehension involves identifying many of the parts of the situation, concentrating on some, and relating them to each other and to scripts. Scripts are apparent in the identification of a pulley and of a demonstration. The details of the man's clothing and the support of the wheel are ignored, as not adding to comprehension of the situation. If a satisfying interpretation had not been reached quickly, perhaps the viewer's attention would have shifted to details that were ignored initially, in a search for clues that would make the situation meaningful. The identification of the scene as a physics demonstration is a generalization about the whole, in this case determining what the purpose behind the scene might be. That classification of the whole seems to be a necessary part of comprehension: people need to label the situation they are in or are viewing.

Another facet in comprehension of situations is being able to explain why the scene is constituted as it is. This may not be as necessary as labelling is, but is a strong component in comprehension. If I am in a car on a freeway, and I find the traffic banks up and eventually stops, I comprehend the

Figure 5.4 Apparatus used in probing understanding through prediction–observation–explanation

situation because I have a script for it and can label it: I call it a traffic jam. My comprehension is increased when I can think of reasons for the jam: I know there are many cars about because it is the time of day when people are on their way home from work in the city centre, and I suppose there has been an accident or a breakdown which is blocking some lanes, or perhaps there are road works. Sometimes a situation can remain puzzling for a while, with comprehension coming later as reasons are thought of, or supplied by someone else. Once, on a train, I sat opposite a young woman and a man who was about 30 years older than her. She was speaking in a pleasant tone to him in a tongue foreign to me, when in the middle of her speech, without a change in her amiable expression, she said in English 'If you don't give me back my flask I shall have to call the police'. The man replied quietly, and the conversation continued. Soon after, they left the train together, apparently quite friendly. I could not comprehend why she said that one English sentence to the man; or why, having made such a threat, she should seem to be on easy terms with him. Hours later I described the scene to a friend, who made it instantly comprehensible by providing an explanation: the woman was recounting an earlier event which had happened to her, one in which the man was not involved, and in which the English sentence had been spoken by her or someone else.

Prediction is a sort of explanation in advance. We feel we comprehend a situation when we know what is likely to happen next. I know the traffic jam will clear in time, and I will pass the cause of the obstruction. I did not comprehend what was happening in the train, and could not predict what would happen. Both explanation and prediction are involved in comprehension of scientific phenomena. The air pump starts and the sound of the bell in the jar decreases. The pump is stopped and the demonstrator moves to turn the tap that lets the air back in. We feel we comprehend this event, because we can explain why the sound decreased and can predict that it will return to its former level when the air goes in.

Measurement of understanding

I have defined understanding of the various targets in terms of internal patterns of knowledge or internal processes of analysis or of forming explanations. Clarity about these unobserved patterns and processes is important, because it makes the nature of learning clearer and because it brings forth implications for teaching. If we know what we mean by understanding, we have some chance of teaching for it effectively.

There is also some value in thinking about the overt consequences of understanding, because they will suggest how we can test whether students have good patterns of knowledge or have comprehended what we tell them. We could go about this by considering each target in turn, but as a single method of testing might be useful for more than one target, it is more convenient to look at each method, discussing its application and strengths and weaknesses, and how it relates to the targets.

Until recently the only format that was available for testing understanding of science was the standard sort of problem that is common in school tests. In the past decade, however, researchers have developed a wider range that includes various forms of interviews, concept maps, word associations, prediction–observation–explanation tasks and Venn diagrams. Each has particular uses.

School tests

The commonest method of assessment of almost anything to do with mental development is the school test. The similarity in style between the tests used in schools throughout the world is remarkable. It is as if there are universal pressures which force the evolution of tests into a limited range of question types. There are objective questions (so-called because of the relative lack of indecision in scoring them, though setting them is subjective) which may be multiple-choice or may require a word or a number as an answer, and extended-answer questions, which require anything from a line to pages of response. Within both forms the recollection of a fact may be called for, or a problem might have to be solved. Recollection is not regarded as a measure of understanding, but solving a problem is.

The analysis of understanding with which this chapter began is not consistent with the view that solution of a problem is a pure measure of understanding. It may involve understanding as I have described it, but it requires more. The extra requirement is a set of cognitive strategies, which I did not include in the earlier discussion because that was about understanding of disciplines and smaller divisions of knowledge, and strategies are broad skills which apply across disciplines. Inability to solve a problem might come about through lack of organized knowledge, that is lack of understanding, or through lack of a cognitive strategy, that is lack of ability. On the other hand, solution of the problem will normally demonstrate understanding, plus ability. It will not, of course, if drill has converted the task from a problem where the procedure has to be worked out to simple recall of the answer or of the algorithm needed to obtain it.

This is not an argument against the use of problems in school tests. Without them tests would revert to the rote recall of facts or algorithms that was common in the 1930s and earlier, and would contain no element of understanding at all. It is perfectly reasonable to use problems as tests of reasoning capacity plus understanding of subject knowledge, which may be the combination of qualities required, say for admission to further study. The point is merely that problems test more than understanding.

Much of the frustration of teachers and students results from failure to distinguish between understanding and the ability to solve problems. Because understanding of a topic is a matter of acquiring propositions, skills, images and episodes, which is not at all difficult, and seeing relations between them, in the course of a few hours teachers can help students attain it to a good degree. But cognitive strategies take a long time to learn, and despite the best efforts of the teacher and the student problems may still be beyond the student's capacity. Failure at problems leaves it obscure whether the difficulty lies in lack of understanding, about which successful action can be undertaken immediately, or lack of strategies, which requires a much longer and more radical treatment. I will discuss that treatment later, in the next chapter; for the moment I will merely comment that schools are not constituted in a way that makes it easy to do much about strategies. The point of this discussion is that, although school tests based on problems are useful, other tests that measure understanding without requiring so much in the way of strategies are also valuable, if only in order to identify the source of a person's failure to perform adequately. Strategies are a relatively minor feature of most of the methods of assessment that follow.

Prediction–observation–explanation

This procedure was devised by Champagne, Klopfer and Anderson (1980) to probe comprehension of a situation. In revealing what is comprehended about a situation, something may also be discovered about understanding of the concepts that are involved. The procedure, which can be used with large groups, is a sharp, powerful means of uncovering deeply held beliefs about scientific principles and phenomena, in contrast to school tests, which often tap only the veneer of facts and algorithms that students have learned in order to get a good grade.

The procedure is based on the classic model of research: a hypothesis is stated and reasons are advanced for why it might be expected to be true, data which bear on it are gathered and the results are discussed. In the investigation of comprehension respondents are confronted with a situation and asked to write their predictions of what is going to happen. For instance, they are shown a block and a bucket hanging side by side on a pulley; the block is pulled down and they are asked what will happen when the block is released (see figure 5.4 above). Reasons for the prediction are requested. The event is then allowed to take place, and the respondents are asked to record what they observed. They are then required to explain any discrepancy between what they predicted and what they saw happen. Using this technique with a variety of tasks, Champagne, Klopfer and Anderson (1980) and Gunstone and White (1981) found that many university physics students in the United States and Australia had views of force, motion and gravity that were at odds with scientists' views, despite their being skilled in the use of physics formulae.

The prediction–observation–explanation technique is readily applied to situations in all sciences, though in biology some of the changes might be slow. That could be overcome with time-lapse photography or through second-hand observations, though much of the value of the technique lies in the presence of a real situation. It is also applicable in non-science disciplines. In mathematics, people could be asked what will happen to the shape of a curve if one of its terms is altered; in history, a real or an imaginary situation could be described and people asked to predict what happened; in music, some bars of Mozart could be given, and the next few bars could be asked for, as a prediction, rather than recall of information.

Venn diagrams

Venn diagrams are a way of exploring the meaning people give to concept labels. Like prediction–observation–explanation, they can be used with large groups, but unlike that technique they are directed at the understanding of relations between a small number of concepts rather than comprehension of a situation. They contrast with most other techniques in requiring a relatively non-verbal response, and so may tap aspects of understanding of concepts which other techniques do not.

The procedure is to ask students to sketch a representation of the relation between several concepts. Some practice may be necessary first to acquaint respondents with the procedure. To test understanding of the terms 'base', 'alkali' and 'hydroxide' one could ask students to write definitions, but a Venn diagram may be a better probe. A respondent might produce the diagram shown in figure 5.5*a* and immediately we can infer that the person believes all alkalis are bases, that some alkalis are hydroxides, and that there are some bases that are neither alkalis nor hydroxides, and so on. The test may be extended by asking for specific instances from different regions of the diagram, and for the region into which a given instance falls.

Figure 5.5*b* shows a science graduate's representation for the set of flowering plants, trees, grasses. In a way the set is not a fair one, because trees are a common class, not a biologist's, while the other two are both common and technical. However, the diagram does show the graduate's

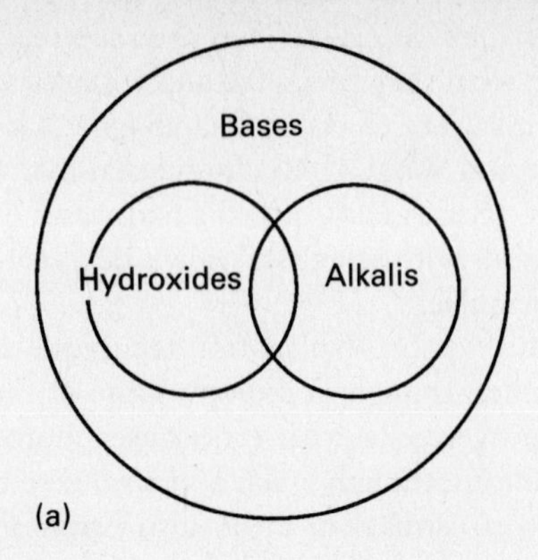

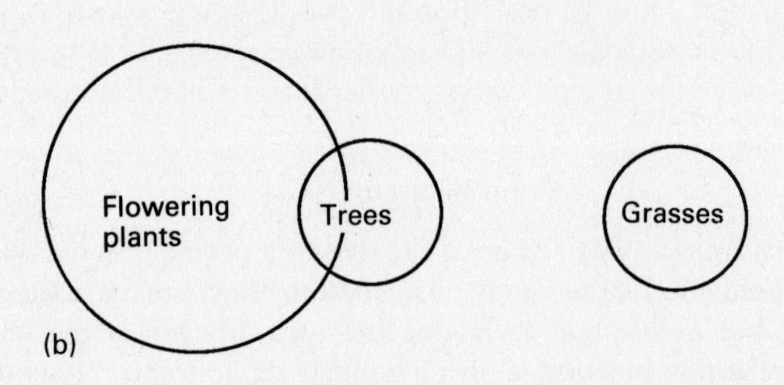

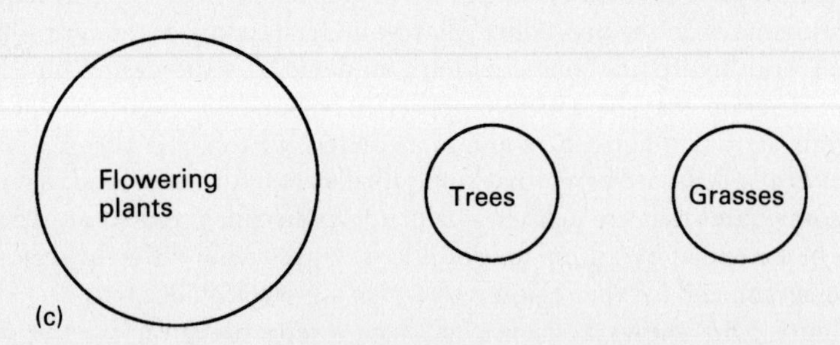

Figure 5.5 Examples of Venn diagrams from (*a*) a hypothetical person; (*b*) a science graduate; (*c*) an eighth grade student

lack of understanding about the nature of grasses, and, less obviously, about the nature of flowering plants. Figure 5.5*c* is the representation of the same set produced by an eighth grade student who reveals a lack of knowledge about trees, too.

Other examples of the sorts of sets this technique might be used with are: objects with zero acceleration, non-zero acceleration, zero velocity, non-zero velocity; animals, mammals, insects, reptiles, amphibians; elements, metals, alloys. Non-science examples are poems, sonnets, odes; tyrants, dictators, demagogues. Further examples are given by Gunstone and White (1986).

Although restricted to only one aspect of understanding, i.e. the relation between one concept and another of a similar type, the Venn diagram technique is powerful and useful. It can be used with groups such as a school class, and is valuable in interviews. Questions can be asked about why the diagram has been drawn so and whether the areas and degrees of overlap correspond to numbers of cases. Sensitivity would then be added to power and usefulness.

Concept maps

Suppose a teacher wants to check on how well students understand the structure of a large body of knowledge, say a term's work on astronomy, ecology or educational psychology. The students could have learned the subject matter as unrelated facts, or, as the teacher probably hopes, as a well-integrated collection of topics. Concept mapping is a practical technique that takes only a short time and can be used with a whole class to check on this aspect of understanding.

The procedure is described in detail by Novak and Gowin (1984), who give many examples of its use. Key terms ('concepts') of the topic are written on cards which the students have to use in a set of operations:

1 Set aside any cards bearing terms that are unfamiliar.
2 Place the cards on a large sheet of paper in an arrangement that makes sense. Closely related terms should be close together.
3 Keep rearranging the cards until you are satisfied with the pattern.
4 Fix the cards to the sheet.
5 Draw lines between related terms.
6 Write on the lines the form of the relation.

In this way a pattern such as that shown in figure 5.6 is obtained.

Variations on the technique are to give the students blank cards on which they first must write their own selection of key terms, or to have the final step carried out by an interviewer who elicits an oral statement of each

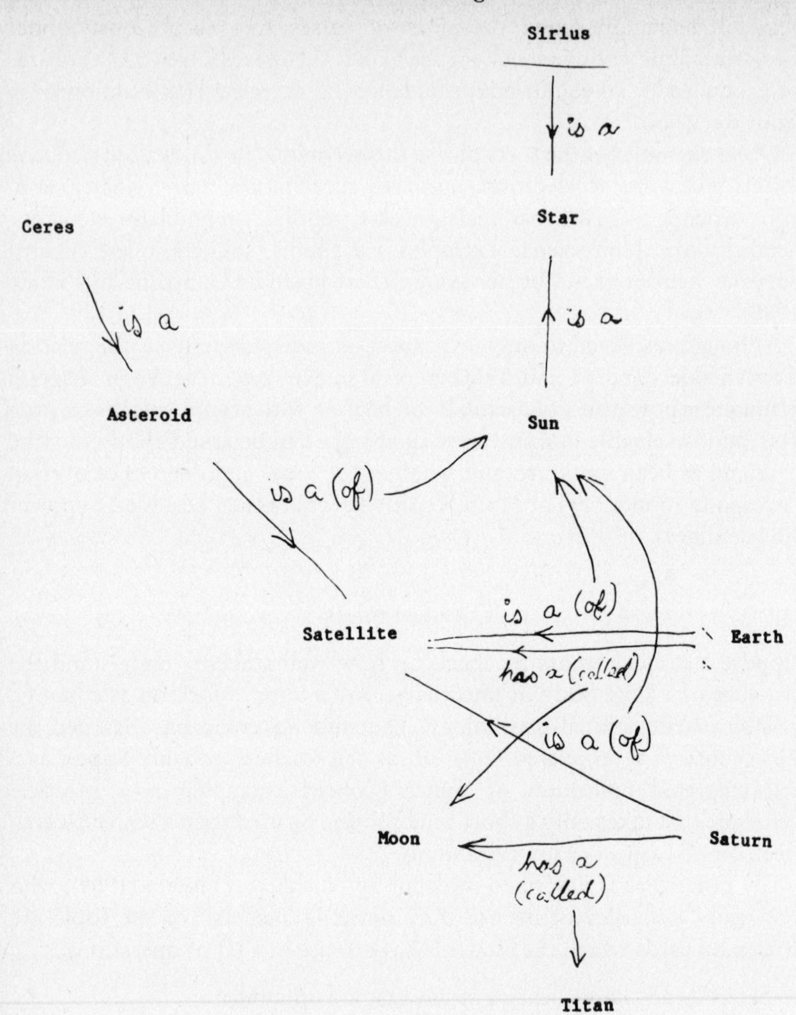

Figure 5.6 A student's concept map

relation from the student (e.g. Champagne et al., 1981). Where the technique is used for training rather than assessment, the teacher may prefer the students to work in small groups rather than individually.

In all variations the essential nature of concept maps remains the production of a pattern of terms linked by relations, a coarse representation of how the respondent sees a substantial body of knowledge. It is a measure of understanding of, if not a whole discipline, at least a large topic within one. This breadth is coupled with a lack of clarity in the measurement.

While the procedure has the advantages of covering large topics, being quick to administer, and capable of being used with large numbers of people at a time, it yields a less sharp assessment of level of understanding than do other methods. The maps are difficult to reduce to a numerical score, if that is desired. The advantages must outweigh the drawbacks, however, for in my experience teachers readily adopt the technique.

Word association

Although word association tests are as easy as concept maps to use with large groups and provide a similarly coarse assessment of understanding, they are not as popular with teachers.

Several varieties of procedure for word association tests have been compared by Shavelson (1974; Shavelson and Stanton, 1975) and Preece (1976a, 1976b); the niceties of their details do not affect the comments that can be made about the general notion. One of the commonest procedures is that used by Johnson (1964) and Shavelson (1972, 1973; Geeslin and Shavelson, 1975). The creator of the test selects a small number, typically about ten, of key terms from the topic. These terms are placed one at a time before respondents (or read out to them), who are asked to write for each terms as many related terms as possible in one minute. Figure 5.7 shows a hypothetical set of responses to three stimulus terms.

Force	*Acceleration*	*Mass*
acceleration	faster	weight
speed	slowing	force
mass	mass	church
movement	time	priest
reaction	velocity	
police	force	
push		
pull		

Figure 5.7 Example of associations with force, acceleration and mass

The responses may be analysed in several ways, most usually by deriving for each person a matrix representing the similarities between the responses to each pair of stimulus terms. In figure 5.7 the terms 'force' and 'acceleration' are more closely related than the other two possible pairs because they contain more terms in common. Shavelson used Garskof and

Houston's formula (1963) for comparing lists, to obtain a quantitative index of relatedness. For lists of this type the formula reduces to:

Relatedness Coefficient = (sum of products of rank orders of words common to both lists)/($\sum_{k=1}^{n} k^2 - 1$), where n is the number of words in the longer list.

This index involves assumptions which may not command general assent. The sequence of responses to a stimulus term is taken to represent the order of closeness of relation that the respondent sees between them. Thus in figure 5.7 'mass' is to be thought of as more related to weight than it is to church. Another assumption is that equal values of the index indicate similar degrees of association.

The two pairs of lists in figure 5.8 have indices of similarity of 1.0, but many people might feel that the second pair reflects greater association. A third assumption is that association with a third term reduces the similarity between a word and another. Thus if the word 'caustic' is added to the first list stimulated by alkali in figure 5.8, the index of association between the first pair of acid and alkali falls from 1.0 to 0.3. There is also the assumption that non-identical words add nothing to the similarity of two lists, so in figure 5.7 speed and velocity are just as different in the force and acceleration lists as words such as police and mass.

Acid	*Alkali*	*Acid*	*Alkali*
alkali	acid	alkali	acid
		litmus	litmus
		pH	pH
		neutralization	neutralization
		salt	salt
		reaction	reaction

Figure 5.8 Two pairs of word associations, with index of similarity of 1.0 in each pair

The assumptions behind the similarity index may make people cautious about its use. However, if it is accepted as a measure of the association the respondent sees between the terms, the matrix of the values of the index can be used to derive a picture of the relations between the terms which will look similar to that obtained in concept mapping. The higher the value of the index, the closer the terms should be placed together. Unfortunately, as Strike and Posner (1976), Stewart (1979) and Sutton (1980) all point out, the nature of the relation the person sees between the terms is absent from the representation. All that is known is that the person associates force and acceleration, not why. Gunstone (1980) amended Shavelson's procedure to remedy this weakness. He required respondents to write a sentence for each stimulus response pair, incorporating both terms.

Even with Gunstone's amendment, the word association procedure provides only a rough measure of understanding. The full richness of the association between terms remains hidden. Thus, although like concept mapping it is a convenient technique, others of greater fineness, aiming at understanding of single concepts or even elements, are needed too. Most of the finer techniques in recent use in science education research involve interviews.

Interviews about a concept

Whereas concept mapping and word association are used with large parts of a discipline, interviews usually deal with single concepts or, finer still, single propositions. The main aim in an interview about a concept is to obtain as complete as possible a set of elements of knowledge that the person relates to the concept's name.

Many procedures are possible for interviews. The one followed will depend on the interviewer's purpose and own model of memory and understanding. For example, Gunstone and I (White and Gunstone, 1980) wanted to assess the understanding that science graduates in a pre-service teacher training course had of two topics – current electricity and eucalypts – in order to see whether they were consistent across topics in their tendencies to inter-relate elements of knowledge and to store images and episodes. We thought that some people would tend to recall lots of images, for instance, while others would be more fact-oriented.

We had an open beginning: 'Tell me, what do you know about electric current?' Most of the science graduates responded freely with statements which we inferred were based on propositions. All that was needed was an occasional nod of encouragement or a request, 'Can you think of anything else?'. In order to elicit episodes and images, when the flow of propositions dried we asked, 'Do you have any personal experiences relating to electric current?', and later, 'Do you have any mental pictures relating to electric current?'. At the time we were concentrating on visual images. We might have asked whether the words 'electric current' brought any sounds, physical feelings, smells or tastes to mind.

The procedure then became increasingly specific. We asked the same three questions about knowledge, experiences and mental pictures for terms that are important in relation to current, such as resistance, potential difference, Ohm's Law, charge, insulators, batteries, and AC and DC. Then we asked what formulae the respondents knew involving current or any of the other terms we had introduced, what definitions, what things they saw as similar or analogous to current, properties of current, its production, its uses, its effects, its measurement, types of electric current, what they knew about its history, and whether names of any people associated with electric current came to mind. In a last attempt to uncover episodes or images, we asked whether they recalled any incidents in films or books in which electric current played a major part. Finally they were asked whether anything else about electric current had occurred to them. Then we repeated the procedure for eucalypts.

Sets of elements such as the one in figure 5.9 were derived from recordings of the interviews. The transfer from tape to written form is essentially verbatim, with hestitations and false starts and the utterances of the interviewer omitted. Some pronouns have been replaced by nouns where it was obvious what they were meant to be.

Many other procedures are possible. The one above brings out knowledge of propositions, images, episodes and strings. It does not recover intellectual and motor skills and cognitive strategies. Nor is it as specific as it might be. We could have phrased it like a school test, by asking direct questions: 'What does a galvanometer do?', 'What is the formula for the effective resistance of a set of resistances in parallel?'. That would have made it more likely that all the respondent's knowledge had been ascertained, but we did not do that because our purpose was to see what knowledge came freely to mind without the cue of a direct question.

Lists such as the one in figure 5.9 are cumbersome, particularly when the person is much more informed about the concept than the one used as an example in the figure, and so a researcher or teacher may need a method for summarizing them. This is a common problem, for we often have complex things to describe in a summary way. The form of the summary depends on the purpose for which it is to be used, but usually it will involve reporting values on anything from one to a large number of dimensions.

If someone asks you what your house is like, you can sketch a plan and give values for area, number of rooms and number of storeys, and for nominal scales such as type of wall material and colour of carpet. The same approach can be taken with the results of an interview. Usually interviewers are interested in how much someone knows, that is the extent of knowledge. They also make judgements about whether the knowledge is correct, that is, whether it agrees with what they believe to be true, and about how precise it is. Often in school practice these properties, as revealed by a test, are coalesced in a single figure – grade B, or 72 per cent for biology, say. It is possible to propose other dimensions that might be important for particular purposes. For example, the proportions of

Electric current is charged particles moving through a medium of some sort.
Electric current goes faster through a conductor.
Metal is a good conductor.
An ionic solid is a good conductor.
Electric current comes out of the plugs.
I see electric current as little electrons running through.
I think of electrons as an *e* with a negative at the top.
I've plugged in plugs.
I associate charge with atoms and ions.
Protons have positive charge.
Electrons have negative charge.
I visualise something out of Stove and Phillips – some diagram showing electrons running through.
Potential difference is the difference of electric charge in two different areas.
Battery is a thing that goes in a car.
Battery is a thing you stick in a torch.
Battery has a positive terminal and a negative terminal.
Battery is usually a square thing.
I used batteries on the teaching round to do an experiment in chemistry.
When the car battery has been flat I've had to put jumper leads on it – twice.
Insulators are used to prevent electric current getting from one place to another.
AC and DC have something to do with electric current.
Resistance is a little squiggly thing.
I suppose we did something with resistances in fifth form.
Ohm's Law has an I in there and a V.
Ohm's Law has something to do with electricity.
Electric current is continuous.
Electric current turns on lights.
Water has something to do with making electric current.
Yallourn power station makes electric current.
Electric current is used in the home.
Electric current is used for light.
Electric current is used to move things in motors.
DC stands for direct current.
Amp has something to do with electric current.
Amp was French.

Figure 5.9 Elements of a science graduate's knowledge of electric current, obtained through interview

different types of element, if these have been revealed by the interview, may tell something about the person's style of learning. By using the two relatively unrelated topics of electric current and eucalypts, Gunstone and I were able to identify people who consistently formed many episodes in relation to both topics, others who tended to have images and yet others who stuck to propositions. There were, of course, all shades in between. There is also the issue of whether the elements a person recalls tend to relate only to topics which are a part of the subject, such as resistance, batteries and voltmeters in the case of electric current, or whether there are others mentioned which appear to have little to do with the subject. Someone who relates electric current not only to integral concepts but to things such as health, insects or cattle has a different style of learning from someone who keeps the knowledge in its own field, and may be said to have a different form of understanding. In *The Act of Creation*, Koestler (1964) suggests that linking diverse topics is an essential part of imaginative solution of problems. This dimension may therefore be one of great importance. Another dimension is the overall linking of the elements of knowledge. For some people knowledge of a concept may be highly integrated, with every element closely linked to others. For other people knowledge of that concept may fall into a number of sections which are not strongly linked with each other. The former arrangement would reflect greater understanding.

No one of these summary measures of extent – accord with authority, precision, the proportions of types of element, the ratio of numbers of internal and external connections, and the pattern of associations of elements – is by itself a sound measure of understanding. Each reflects an aspect of the knowledge the person has about the concept and hence about the person's understanding of it. The point is that understanding is not identical with any one of the measures. Understanding is a complex, multifactor notion which cannot be described simply.

Interviews about situations

The procedure followed by Gunstone and myself, of simply asking people what they know, is open to the objection that such an unfocused task may result in the respondent forgetting about a whole section of knowledge. We tried to circumvent that by moving from the open initial request to questions about terms that might be related to the concept. Another way of giving an interview a focus is to confront the respondent with a situation, which often will involve a task, and use the situation to stimulate a conversation that will elicit beliefs about the concept of interest.

Interviewing based on a task is one of the inheritances science education research received from Piaget. A comparison of the reviews of Shulman and Tamir (1973) and White and Tisher (1986) reveals that there was a sharp increase during the 1970s in the appreciation of Piaget's work, and although most attention was given to his notion of developmental stages, impetus was also given to interviewing as a means of assessment. The application of interviews was broadened: where Piaget was concerned with achievement of general conservation skills and logical relations, the varieties of interview described here are directed at the respondent's understanding of a single concept.

Osborne and Gilbert (1980) used questions about situations represented in simple line diagrams to elicit respondents' beliefs about the concepts of work and electric current. Examples of their diagrams and questions are shown in figure 5.10.

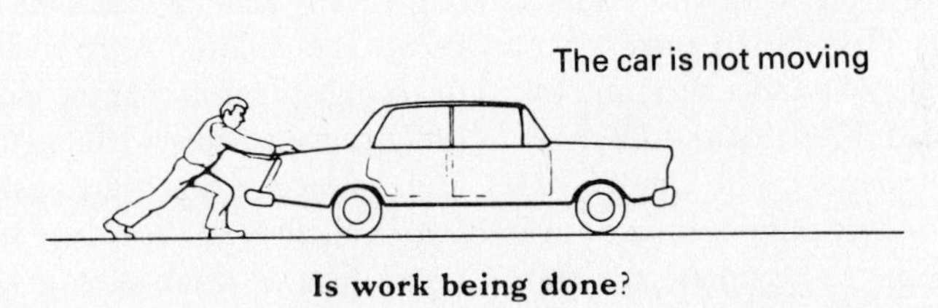

Is work being done?

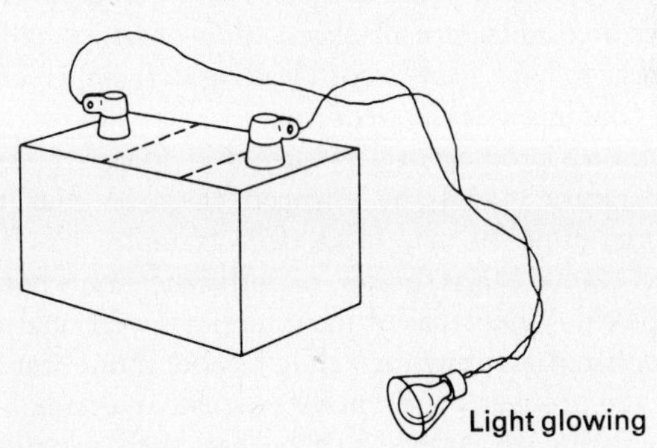

Is there an electric current in the battery?

Figure 5.10 Line drawings used by Osborne and Gilbert (1980) in interviews about instances probing understanding of work and electric current

Osborne and Gilbert named their procedure 'interview about instances'. It has been used extensively at the University of Waikato Learning in Science Project to assess understanding of friction (K. Stead and Osborne, 1981), light (B. F. Stead and Osborne, 1980), force (Osborne, 1980) and other concepts. The statements respondents utter in the conversations are used to make inferences about their understanding of the concept. There is no regular procedure for this, but in practice it has not been found difficult to identify certain recurring beliefs. The technique revealed Aristotelian views of force and motion, and images of electricity as a one-way, non-returning, flow.

Other interviewers have used actual equipment rather than line diagrams, though the course of their conversations is much the same as that in interviews about instances. Examples are studies at the University of Paris (e.g. Guesne, Tiberghien and Delacôte, 1978) and Erickson's investigation of the concepts of heat and temperature (1979). Among the classic Paris studies is a probe of the understanding seven- and 12-year-olds had of electricity (Tiberghien and Delacôte, 1976). The children were given a dry cell, torch globe and a wire, and asked to arrange them so that the globe was illuminated. Fredette and Lochhead (1980) repeated this with engineering college students in the United States, and Andersson and Kärrqvist (1979) with secondary school students in Sweden. In all three studies many students were unable to complete the task. Observation of their attempts, rather than their spoken comments, was the main source of information about their concept of current. A common arrangement was to place one end of the wire on the positive terminal of the cell and the other end on the base of the globe. This could be interpreted as a one-way flow view of current, which may be held even by people who have the proposition 'current flows in a circuit'. Andersson and Kärrqvist make the interesting suggestion that the view is a consequence of observation of lamps and other appliances, which appear to have a single cord leading to them which most people may not know contains at least two wires.

In Erickson's probing of 12-year-old children's understanding of heat and temperature (1979), he engaged them in four tasks to provoke conversation. One of the tasks, for example, was to place different substances – metal, sugar, butter, naphthalene – on a hot plate and observe what happened. Videotapes of the interviews were analysed to derive a set of statements representing each child's belief about heat and temperature.

One of the problems with interviews is that although they are sensitive ways of eliciting an individual's beliefs, they are time-consuming and usually so few people can be studied that generalizations about a population are not possible. Erickson combined the sensitivity of interviews with the power of mass tests by deriving from his transcripts three different views of heat and temperature which covered most children's explanations of the phenomena they had observed. These were kinetic and caloric views, and a set of ideas which he named 'children's viewpoint'. Erickson developed a multiple choice test, in which phenomena were demonstrated and described, and the children had to select which explanation of three, representing the three views, was the best on grounds of agreement, clarity, ease, truth, familiarity and similarity to own ideas. On analysis, Erickson found that the two factors of belief and familiarity were behind the children's evaluations of the explanations. Erickson used his test with grades 5, 7 and 9. He found that some children accepted both the kinetic and caloric explanations, which he interpreted as revealing a particularate view of the physical world with no discrimination between matter and heat. Others rejected the kinetic view, preferring the caloric and children's explanations, which Erickson interpreted as a common sense or intuitive view. Presumably that is one based on an interpretation of heat phenomena which is derived from analogy with observations of other physical events, such as flow of water.

Interviews about single propositions

Nussbaum and Novak (1976) provide the classic instance of the use of interviews to probe understanding of a single proposition. They employed a sequence of questions to identify five levels of understanding that second grade children had for the proposition 'The Earth is round'.

Most of the tasks Nussbaum and Novak used required the children to predict a direction of fall. The interviews began with questions such as 'What is the shape of the Earth?', 'How do you know it is round?', and 'Which direction would you have to look to see the Earth?'. Then the children were shown a globe with a figure of a girl stuck on it. The figure was placed at different places on the globe, and for each the students were asked to draw on a corresponding two-dimensional diagram the direction in which a rock the girl was holding would fall. Figure 5.11 shows three

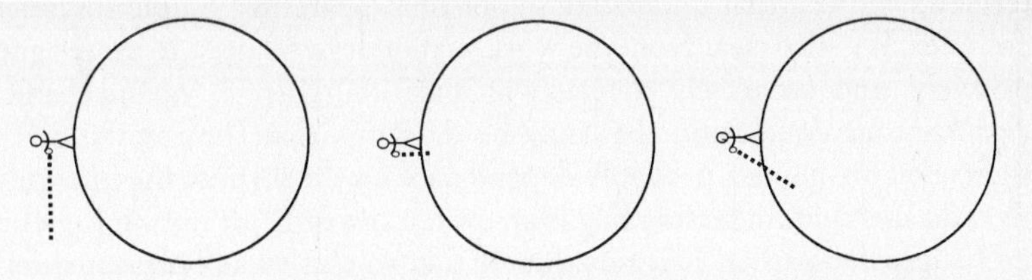

Figure 5.11 Responses of children observed by Nussbaum and Novak (1976) to a question about the direction in which a rock would fall when dropped at a point on a model of the Earth (copyright © John Wiley and Sons Inc.)

responses which were observed. The right hand example is particularly interesting, interpreted by Nussbaum and Novak as a compromise between egocentric and scientific appreciations.

Further tasks in the same interview involved bottles of water placed at different points on the globe, children hanging from swings, and the path a falling rock would take if dropped down a deep hole. All of the tasks were designed to probe deeper and deeper into the meaning the children had for the proposition that the Earth is round.

The five notions that Nussbaum and Novak identified are:

1 'The Earth is round' refers to the roundness of hills on the sky or roads; the ground we stand on is flat.
2 The Earth is round like a ball, but there is a general downwards direction towards a flat ground or ocean somewhere 'below' the South Pole.
3 Similar to the second level, without the ground or ocean so that the Earth is isolated in space, but things in the Southern hemisphere still fall off into the sky.
4 Nearly the scientific view, but up and down are not related to the Earth's centre.
5 Appreciation that the Earth is spherical, surrounded by space, with things falling towards its centre.

An interview like this is, of course, time-consuming, and for practical reasons it may be necessary to surrender some of its sensitivity in return for ease of administration to groups. Nussbaum (1979) did this by producing a multiple-choice test based on the five notions that he and Novak had found. Erickson, whose interviews on heat and temperature were described above in the section on interviews about situations, also followed this procedure (Erickson, 1979, 1980).

Virtually all of the children whom Nussbaum and Novak interviewed subscribed to the proposition that 'The Earth is round'. However, the interviews showed that understanding of this apparently simple statement is complex. It is also clear from the work that interviews need to be designed sensitively, and separately for each proposition; there is no mechanical form. Another conclusion that may be drawn is that the results, or the insights that are gained, probably depend a lot on the form of the interview. This indicates that understanding is an even more difficult notion to define than I suggested earlier. It is remarkable that so elusive a construct should hold such a central place in the aims of education and should be so commonly used in everyday speech.

We have now considered interviews on concepts, situations and single propositions. In all cases we cannot be sure that what people choose to tell us is what they really know or believe. Sometimes the information may not come to mind, or the respondent may filter things out in the belief that they are not what the interviewer is interested in. All that the interviewer can do is be careful to establish rapport and be encouraging. The subjective nature of interviews, the effect of the degree of rapport and the influence of context on what the respondent selects to tell, make interviewing particularly vulnerable as a reliable procedure for uncovering a pattern of associations. Nevertheless, it is an important technique that has revealed gross misconceptions and lack of understanding as well as subtle differences between students' and teachers' views. If it were used more widely and systematically in schools, despite the difficulty teachers have in finding the necessary time to spend with single students, it would yield deeper appreciation of the quality of students' learning and would be likely to lead in consequence to adoption of more effective methods of teaching.

Studies of understanding of science

Researchers have been active in using the foregoing techniques to probe the understanding of scientific concepts and propositions by students in primary, secondary and tertiary levels of education. In addition to the studies I have cited, many more examples are described by Archenhold et al. (1980), Driver (1983), Driver, Guesne and Tiberghien (1985), Duit, Jung and von Rhöneck (1985), Helm and Novak (1983), Osborne and Freyberg (1985), and West and Pines (1985). The studies show that students often have well established conceptions that differ from those of scientists and that persist despite the efforts of their teachers. It is apparent from this work that learning is more than simple addition of propositions to memory, and that teaching for deep understanding is a more difficult and subtle task than merely setting out facts clearly and checking that they have been absorbed. Chapters 9 and 10 return to these matters.

6

Abilities

Ability, like understanding, is a word that is often in the mouths not only of psychologists and academic educationists, but also of teachers, parents and students. All speak of it as a factor that influences learning. But again like understanding, its meaning is by no means fixed and uncontroversial. For some, ability is an unvarying characteristic, for the most part determined genetically or, more fundamentally, to be regarded as a talent given by God. And so we hear, 'Mary is a bright girl who will do well at school', or 'Poor Alan is so stupid, no one will ever be able to teach him anything'. Then for others it is malleable, perhaps varying with content or context: 'Fred might have difficulty with words, but he certainly has a head for figures'. Nor are psychologists themselves of one mind. Wagner and Sternberg (1984) review three conceptions of ability, namely intelligence, level of operational thought and cognitive strategies. In this chapter I concentrate on the last of these, since it is the most optimistic, offering the possibility of increasing the ability to learn, whereas the other two treat it as a less changeable attribute.

All three conceptions refer to general processes. To be 'able' implies a capacity to do things, and can be applied to specific instances such as the ability to hammer a nail, solve quadratic equations, or recite *Hiawatha*. These examples are not, however, the sort of capacity that the term is meant to cover, for they depend on specific elements of memory, respectively a motor skill, an intellectual skill and a string. Rather, ability is meant to be a general capacity, a trait or set of traits enabling one to cope with a wide range of tasks. Prominent among these tasks is the learning of knowledge in such a way that it can be applied to attain goals in situations that are not familiar and routine. Gold (1975) had remarkable success in training severely retarded people to perform complex tasks such as assembling circuit boards or the rear hub brakes of bicycles, by analysing the tasks into discrete operations and encouraging the learners to 'try another way' until they succeeded at each one. Eventually practice smoothed the sequence of operations and the retarded people assembled the circuit boards or hub brakes with facility. They were, however, still retarded. Although they demonstrated some ability in acquiring the complex motor skill, they were hardly any more able than before to cope with new situations on their own.

It is not too fanciful to regard the hub brake as merely an extreme example of what happens in schools. Teachers analyse topics, breaking them into sections that they judge their pupils can assimilate, and try, by giving as much help as they think necessary, to ensure that the students master each section. Of course this is a sensible practice, since the acquisition of elements of knowledge is an aim of all education. It does not, however, do much to make the students more able in the general sense. Certainly acquisition of facts and skills and scripts makes a person better prepared to cope with new situations, because some of the knowledge will be relevant and also greater knowledge may increase confidence and willingness to try. But the term 'ability' has long meant something other than that, something beyond what can be learned through simple instructions. Crucial questions for educators are, what does that general ability consist of, and can it be fostered? The answers a society accepts influence its schools and the nature of the teaching and learning within them.

Intelligence

For most of this century 'intelligence' has been a synonym for general intellectual ability, despite the theoretical difficulties and measurement controversies that have been associated with it. The intelligence view still dominates schools and society, to such an extent that probably the majority of people in developed countries have taken one or more intelligence tests, and many have had the course of their lives affected by them.

The notion of IQ – intelligence quotient – has taken firm root. Many schools keep records of their students' IQ scores, and use them for selection and streaming or for interpretation of performance. Although much less now than formerly, intelligence tests were used for a deliberate social purpose in England and Wales in the 11-plus examination to divide children between the grammar, technical and secondary modern schools. They are often used in determining entry to universities and colleges, while at the other end the IQ scale is used to make judgements about the mentally handicapped, sorting them into the educable mentally retarded and lower categories, which have labels that have become pejorative terms – imbecile, idiot and moron.

Most histories of the concept of intelligence describe its development from Galton's measures of the acuity of sensory perception, through Binet's purpose of identifying those who could benefit most readily from formal schooling, to the selection and placement of large numbers of servicemen in the two world wars. They deal also with the various controversies, such as that between Spearman and Thurstone about the presence of a single general factor, g, or its conceptual replacement by a set of factors. Those with a taste for the more heated aspects of social science may wish to follow the furore aroused by Jensen's interpretation of his research (1969) that in America people of Oriental and European descent are more able than those of African, or Kamin's (1974) and Hearnshaw's (1979) demonstration that Burt's studies of the inheritance of intelligence (1966) were based on invented data.

The controversies, the effort lavished on measurement, the furore, the possible fraud, all show that people take intelligence very seriously. And yet it is by no means certain that the construct is a useful one in education. Under all its versions it presents at least the major part, possibly all, of ability as fixed, a stable characteristic of the individual that cannot readily be improved. In this view, therefore, ability is a factor to be taken into account in teaching and learning, but one not to be meddled with. Its value is in sorting people, not improving them. The implication is that a failure to learn may have been inevitable, and cannot be blamed on the teachers or the school, or really even the student. It is a fatalistic, not an interventionistic, notion of ability.

Level of operational thought

In the early intelligence tests Binet and his co-worker Simon had been interested not only in total scores but also the details of people's performances – their wrong answers and why they had given them. There was no time for analysis, however, when the First World War demanded a method for sorting people into groups that could be trained for simpler or more complex tasks. The vast number of people involved, both as subjects and assessors, meant that the method had to be simple and robust. Reasons for wrong answers could not be probed. Following the war, intellectual testing in the United States and Britain continued along the path that was set during it, of assigning a numerical score based on the number of items answered correctly in a fixed time. But a young associate of Simon, Jean Piaget, a Swiss, had not been caught up by the war nor by the subsequent developments in the English-speaking countries. He continued to probe reasons for wrong answers, and found that often they were not erratic but were developed from a consistent view of the world.

Piaget spent a lot of time in individual interviews with children, studying their development of language and thought, especially their conceptions of causality, how they constructed a picture of the world, their methods of classification, their abilities to conserve quantities such as volume and weight and to arrange objects in order of size, and their conceptions of movement, speed, time and space. From this immensely rich body of work he derived the notion of an intellectual operation:

> Psychologically, operations are actions which are internalizable, reversible, and coordinated into systems characterized by laws which apply to the system as a whole. They are actions, since they are carried out on objects before being performed on symbols. They are internalizable, since they can also be carried out in thought without losing their original character of actions. They are reversible as against simple actions which are irreversible. In this way, the operation of combining can be inverted immediately into the operation of dissociating, whereas the act of writing from left to right cannot be inverted to one of writing from right to left without a new habit being acquired differing from the first. Finally, since operations do not exist in isolation they are connected in the form of *structured wholes*. Thus, the construction of a class implies a classificatory system and the construction of an asymmetrical transitive relation, a system of serial relations, etc. The construction of the number system similarly presupposes an understanding of the numerical succession: n + 1. (Piaget, 1953, p. 8)

Piaget inferred from his research that operational thinking develops in all people in a sequence of stages, which he described in his book *Logic and Psychology* (1953). In the first stage following birth, the sensorimotor period, the child can display intelligent behaviour but no operational thinking takes place. For instance, only gradually the infant realizes that an object continues to exist when it is hidden by a screen, and can be recovered by reversing the action, by removing the screen. When language, mental imagery and symbolic play emerge, the child has entered the stage of preoperational thought. However,the child finds it much more difficult to think about an action than to perform it. When asked whether someone else can see an object from a different viewpoint, the child would have to go to that point before being able to answer. Conservation is not yet developed, so that when a plastic ball is flattened into a different shape the child thinks there is a different amount of the substance, or when a third object is placed between the two others believes that the distance between the original pair changes. Full reversibility of actions, and the notion of conservation of quantities, marks the attainment of the concrete operations stage. However, the child cannot yet think abstractly, through hypotheses. That does not occur until the final stage of formal operations is reached, when instead of thought having to proceed from actual objects to the theoretical, it can now begin with theory. Propositional operations, of the logical forms if this then that, if not this then not that, and so forth, become possible.

Piaget asserted that the order of acquisition of the stages is the same for everyone; that structures formed in one stage are integral parts of the next; that stages are not just a collection of unrelated skills but require a synthesis of them into a new coordinated whole and that there will be development within a stage from preparation for the new level of thinking to the completion of the synthesis when equilibrium is attained.

The notion of stages was picked up by science educators and came to dominate research in the 1970s. Much of this research, however, was not illuminating, consisting of attempts to replace Piaget's careful individual interviews with mass-administered pencil and paper tests and to use these to measure proportions of groups in each stage or the correlation of stage memberships with intelligence test scores or school achievement (White and Tisher, 1986).

The quality of the research does not matter as much as the conception of ability that the stage notion promotes. It suggests that ability is determined, even represented, by one's current stage of development. Thus although not fixed, since one does in fact develop within each stage and does move from stage to stage, ability is not readily improved by direct instruction. Development occurs through physical growth and a combination of physical and intellectual interaction with the world. Experiences have to be interpreted and assimilated so that gradually one becomes able to think in more abstract ways.

While these insights of Piaget are powerful, it is unfortunately easy to convert them to a rigid system in which people are thought to belong to particular stages at particular ages, and in which no concern need be given to supplying them with experiences that may help them form advanced ways of thinking. It can be interpreted as a prescription against exposing young children to abstractions, and so can lead to an impoverished curriculum. It can also lead to ignoring of differences between individuals of similar ages, since they are often assumed to be at the same stage.

These interpretations of Piaget's theory have implications for schools, especially for curricula. Topics were chosen, and units developed, for the Australian Science Education Project (ASEP) in accord with a view that many children in lower secondary school would not have attained Piaget's stage of formal operations. Shayer and Adey's analysis of science topics in terms of demand in operational level (1981) has had wide currency in Britain. The stage part of Piaget's theory causes no great difficulty for schools, for it suggests only that some topics should be left till later. It is also a convenient explanation for failure: 'The children did not learn, not because I taught them badly or because they did not try but because they were not at the appropriate stage.' Thus responsibility for learning is removed from both teacher and student, and there is no incentive to analyse the failure further in order to do better next time. Like the psychometric notion of intelligence, this common interpretation of Piaget's stage conception of ability is fatalistic and not interventionistic. It describes ability, but does not tell how to add to it.

Cognitive strategies

The third way of thinking about ability is to conceive of it as a set of identifiable and learnable skills, each of which can be applied to a range of tasks. These are the cognitive strategies that were introduced in chapter 3 as another kind of memory element. Unlike propositions and intellectual skills, they are not subject-specific. They include powerful procedures like generalizing, keeping mind on the task, determining the goal, reflecting, weighing alternatives, and so on. These cognitive strategies determine the quality of performance in learning, in problem-solving and in day-to-day living. Of course, they operate in conjunction with specific knowledge – there has to be a task, some information to absorb, a problem to solve, on which the strategies can be applied.

Although it is a long time since Gagné distinguished cognitive strategies from other types of memory element (1972), they have proved difficult to describe in detail and to study. Theorists' descriptions of them differ, and no one can yet be confident of having set out a definitive and comprehensive list of strategies. Despite the difficulties, the notion of strategies is attractive because it is more encouraging about the chances of improving people's abilities than either the intelligence or the operational thinking conceptions.

In my attempt to describe strategies I group them under three headings: those to do with assessing the situation, those involved in planning what to do next,and those to do with processing information. This grouping is not entirely satisfactory, because it implies that there are three distinct and sequential stages at which quite different strategies are applied. Learning and problem-solving are much more complicated than that, so the grouping is only a convenience for description: the strategies listed in each group are applied in rapid and mixed succession.

Assessing

To assess a situation is to make it meaningful. Perhaps the most important step in coming to terms with any situation, whether to do with learning or not, is to get clear what the goal is, what it is that you want to do or that someone expects you to do.

Schools are safe, protected places for students, as are hospitals, asylums and prisons for their inmates, because most goals, at least the short-term ones, are determined for them. Students are socialized into thinking it unfair if goals in school tasks are not provided. In general life, however, goals are often not stated, and it is up to each of us to form them. Sometimes the goal is easy to determine. If you are going to work and your car will not start, your goal is to get the car moving. That is only the immediate goal in a longer chain, however. The next goal is to get to work, and that in turn is linked to deeper motives, goals of survival through earning money for food and shelter, and maintenance of personality through a sense of community with fellows and of the value of the job to society. As far as the learning of science is concerned, we may not appear to have to worry about this strategy in schools as long as the present system, in which students have little control over the course of their own learning, continues. The strategy will, however, become important if learners take on responsibility for their own learning, deciding what to learn and how. In learning of science outside a formal school system, the strategy determines whether a conscious choice will be made to learn or not to learn. If the strategy is not possessed or not activated, the individual is at the mercy of events rather than in control of them, and no initiative in learning will be taken.

Even though in the present form of schools students expect teachers to make goals clear, and teachers try to do so, learning often goes astray because students lack the strategy of determining the goal. In learning, whether alone with a text or in a class lesson, the strategy involves students in knowing what the topic is and being clear about the task.

It may seem ridiculous to suggest that anyone could sit through a lesson without knowing what the topic is, but Baird (1984) and Tasker (1981) have found that secondary school students are often vague about the topic, and about its connection with earlier lessons:

Teacher to Observer: This is the third lesson in a series aimed at developing a particle idea and going over the states of matter . . . the first one [lesson] we did was also to do with this. It was to do with expansion and contraction of substances – today was a direct follow-on from that.

Observer to Pupil: What was today's one all about?

Pupil: What crystals can do . . . we get a card every so often [a

	work card] and we copy from the card. Do an experiment, maybe with crystals.
Observer:	Does today's work have anything to do with this other work you've been doing? (pointing to the pupil's open exercise book which was showing notes headed ... 'dilution of a potassium permangamate crystal' and 'heating and expanding liquids').
Pupil:	No, not really ... no.

(Tasker, 1981, p. 34)

Even when the topic is written on the board, knowledge of it may go no deeper than recollection of the name without understanding of what it represents. This was demonstrated by a seventh-grade class in a lesson I observed on taxonomic classification. Beforehand the teacher had written 'Ways of Grouping' in large letters on the board, and underneath that 'There are many different ways of putting things in groups'. The lesson began with a clear introduction in which several children were asked, one after the other, to arrange a dozen objects into two groups. The rest of the class were allowed to comment on the suitability of each grouping. Then the class was broken into twos and threes, each of which had about one hundred pieces of plastic to sort into groups; then each group was to be sorted into further groups, and so on. The children did this enthusiastically, but when I asked them what they were supposed to be learning they could not say. They were following directions without knowing what the topic was. As a ninth-grade student said on another occasion: 'I knew what I was doing, but I was not entirely sure why I was doing it' (Baird, 1984, p. 259).

At least the seventh-grade students knew that the task was to sort shapes into groups, even if they did not know what they were supposed to learn. Often students do not apply the strategy of determining the task, as the following example shows. A ninth grade class was following a lesson in which they had established a heating curve for tapwater by doing the same thing for seawater:

Observer:	(joining a group of three boys) This looks interesting.
Pupil:	Yeah, we have to ... fill the beaker to 150 ... then ... get 5 ml of salt, then we put it in here (indicating the beaker of water) and then we put it on there (touching the gauze on the tripod) – then we have to take the heat.
Observer:	I see, and what's it all about?
Pupil:	I dunno – we just have to do it and get the graph.

(Tasker, 1981, p. 34)

Baird (1984) observed a series of seven lessons in which a ninth-grade class was supposed to identify the factors that affect the period of a pendulum. His recordings of conversations discover a shallow understanding of the task. This is from the third lesson:

Observer:	What is the topic?
Student C:	Pendulums.
Student A:	To work out the pendulum, or whatever it's called.
Student D:	Standard form and scientific figures.
Observer:	(to the teacher) What is the topic?
Teacher:	Measurement, and design of experiments.
Observer:	Would that have occurred to any of you?
Students:	No.
Observer:	(to student D) What is the task?
Student D:	To find out different things about pendulums – to find out if weight affects it, how many oscillations, or if the temperature affects it.
Observer:	Why are you doing it?
Student D:	(No answer.)
Observer:	(to student C) What do you think the task is?
Student C:	Find out what moves the pendulum, that is, if wind affects it –just see if anything changes the swing of it – breeze or something.

(Baird, 1984, p. 298)

A lot of the students' difficulty with the task stemmed from not knowing what 'period' meant. If they had applied the strategy of determining the task they would have been more clear than this:

Observer:	You haven't used the word 'period' anywhere, have you – what is your definition of the period of the pendulum?
Student R:	The time of the oscillations.
Observer:	So, what's the period in your results?
Student R:	Twenty seconds (the time over which they had counted oscillations).
Observer:	So, if you timed it for sixty seconds, what would the period be?
Student R:	A minute.
Observer:	(to other students) What do you think the period is?
Student F:	The number of times it goes back and forward.
Student H:	The time till it stops swinging.

(Baird, 1984, p. 229)

The two strategies, of determining the topic and the task, operate through self-directed questions such as What is all this about? What does it involve? What does it relate to? What am I required to do? Does it seem reasonable in terms of the information given? (Baird, 1984, p. 100). People for whom the strategies have become almost automatic may hardly be conscious that they are asking themselves these questions in order to direct their learning. They could, however, be applied deliberately, much as a beginning golfer keeps repeating 'keep left arm straight, eye on the ball, back slow ...'. In time the operation becomes smoothed and automatic.

Until it does, though, students would benefit from training in asking themselves the questions as a conscious act.

As well as determining the goal, one has to make sense of the wealth of information that is available in any situation. Strategies that help in this are those of grouping the information into a manageable number of chunks and of then assessing which of the chunks are important enough to attend to.

Chunking can be unconscious, as when we combine printed letters into words or look at a scene and see it as a collection of familiar objects, or deliberate, as when we break a complex problem into distinct parts. The perception of a scene as containing separate objects is an important stage in comprehending it, as instead of an enormous number of stimuli the mind perceives a limited number of entities. As I look out the window I chunk the scene into some trees, a road, a wall, a man running, a car, the sky, a building, and not either a near-infinite number of separate tiny specks or a continuous undifferentiated smear. Thus seeing involves synthesis. Analysis is also possible, for if I concentrate on any one of the parts of the scene I can divide it into smaller chunks: the car has wheels, windows, lights, doors. That introduces a second strategy, selecting chunks for attention, together with its negative of screening out stimuli from attention.

The chemistry problem in figure 6.1 illustrates deliberate chunking and selecting. Its statement is too large for ready comprehension. By reading it through and thinking about it, a person might divide it into four chunks:

Fuels burn, producing SO_2.
SO_2 is absorbed by $CaCO_3$.
Leftover SO_2 reacts with dichromate.
Leftover dichromate reacts with ferrous ions.

Once these chunks have been formed, it is possible to strip away the irrelevancies about fossil fuels and pollutants and so on.

Selecting the chunks that matter is a key strategy that does much to give able learners their advantage over the less able. A person who lacks it has the difficult task of trying to remember everything, and consequently is in danger of being overwhelmed by a mass of information, while the more able have sorted out the relatively few propositions that matter:

A lot of times I get tied up with all the details, and sometimes I should just leave those out and just worry about the main ideas, but I don't tend to do that. (College student, interviewed by Baird, 1984, p. 79)

Chunking and selecting for attention involve knowledge. The strategies by themselves are not enough, for the person must know facts and skills in order to be able to form useful chunks and to make an appropriate selection of useful information. Knowledge is involved also in the next strategy, of

Many fossil fuels (coal and oil) contain a small proportion of sulphur, and when the fuel is burnt sulphur dioxide (SO_2) is formed and released into the atmosphere along with other products of combustion. Since SO_2 in the atmosphere is regarded as an undesirable pollutant, it is becoming necessary to take steps to eliminate SO_2 from the waste gases arising from the combustion of sulphur-containing fuels. One process being developed to reduce the SO_2 output involves the introduction of limestone, $CaCO_3$, into the combustion chamber where the SO_2 can react according to the equation

$$CaCO_3(s) + SO_2(g) \rightarrow CaSO_3(s) + CO_2(g).$$

In a laboratory test of the efficiency of this process, 1.00g of a fuel containing 0.01g of sulphur in a combined form was burnt in a small combustion chamber containing $CaCO_3$, in a stream of air. The products of combustion were then freed of any dust and bubbled through 10.0 cm^3 of an acidified 0.0100M $K_2Cr_2O_7$ solution, in which all the SO_2 in the gas stream was absorbed according to the equation:

$$Cr_2O_7^{2-} + 3SO_2 + +H^+ \rightarrow 2Cr^{3+} + 3SO_4^{2-} + H_2O.$$

When all the fuel had burnt, and all the resulting gases had passed through the $Cr_2O_7^{2-}$ solution, the remaining unreduced dichromate was titrated with 0.0200M Fe^{2+} solution. It was found that 15.0 cm^3 of the 0.0200M Fe^{2+} solution was required to just reduce the remaining dichromate, according to the equation:

$$Cr_2O_7^{2-} + 6Fe^{2+} + 14H^+ \rightarrow 2Cr^{3+} + 6Fe^{3+} + 7H_2O.$$

Calculate the *percentage of the sulphur* in the original fuel sample which had escaped as SO_2 with the gases leaving the fuel chamber.

(Atomic weight: S = 32)

Figure 6.1 A complex chemistry problem

determining the familiarity of a situation and evaluating one's knowledge and feelings about it: How much do I know about this? Enough to be able to understand it? How important is this to me? How interesting is it? Another strategy is to generate expectations of what is involved and the likely outcome: How long will this take? How hard will I find it? How much effort should I put in? Am I likely to complete the task satisfactorily? These questions are important in preparing the learner to commit appropriate effort.

Unless these strategies are applied, all tasks are approached in much the same way and learning can drift astray. For example, Baird (1984, p. 362) reports students' difficulty in interpreting a graph showing changes in

Australian population numbers since 1788, the beginning of European settlement. They failed to gain much from 20 minutes' application because, as was subsequently discovered, only one student of 25 knew what had happened in Australia in 1788. None had thought to ask about the significance of 1788. Evaluating knowledge can alert people to the need to find out some specific information before learning can proceed well. It can also remind learners of lots of relevant things they already know, with which the new knowledge is to be linked. Evaluating feelings may reveal to you that although you do not, for instance, have much interest in the topic, you do appreciate its importance in obtaining a good understanding of the broader subject. That appreciation will affect the extent and style of processing that you apply to the information:

> I tend to think that my approach is different to these liked and disliked areas. I feel that I have to motivate myself, and struggle more when learning a disliked subject. I have to concentrate harder, read more for less liked sections. I also have to put in more work to less liked sections to achieve the same results. (College student, interviewed by Baird, 1984, p. 80)

Determining familiarity extends to more than deciding whether one had been in that exact situation before – seen that house, that person, that piece of text, that problem. In chapter 3 I described generalized episodes, or scripts. Assessing a situation involves determining whether one has a script for it, in which case it would be highly comprehensible; or whether one has a script that might be adapted to it, providing a lesser degree of meaningfulness; or whether none of one's scripts has any bearing on the situation at all, in which case the situation would be incomprehensible. In reality total incomprehension may never occur: one can always force some meaning into a situation. People may differ in their capacities to do this, which amounts to another strategy, distinct from evaluating familiarity, of establishing or inventing parallels between the present situation and one or more of the individual's repertoire of scripts. There are then two factors which determine whether a person is at a loss in a situation: the range of scripts the person has, and the command of this strategy of seeing parallels.

Seeing parallels is connected with another strategy in assessing a situation, that of perceiving alternative interpretations, which might be described as 'searching for ambiguity'. This is a vital strategy for coping with situations that are not turning out as one expects. Re-evaluation of the original situation may be necessary, and if it can be seen in only one way then all that may be possible is to repeat the previous actions which have already led to an unwanted state. This commonly happens in problem-solving. If the individual has misinterpreted the conditions then failure will cause reversion to the start. If a new interpretation can be made, a new line of attack is possible and the problem might be solved. Otherwise there is no hope for progress and either the problem is abandoned or the same cycle of unsuccessful operations is run through again and again.

Ambiguity causes difficulty in learning as well as in problem-solving. Many misunderstandings occur because people do not check for alternative interpretations:

> 'I mean,' said Father Brown, 'that it's always happening; and really, I don't know why. I always try to say what I mean. But everybody else means such a lot by what I say.'
>
> 'What in the world is the matter now?' cried Greenwood, suddenly exasperated.
>
> 'Well, I say things,' said Father Brown in a weak voice, which could alone convey the weakness of the words. 'I say things, but everybody seems to know they mean more than they say. Once I saw a broken mirror and said "Something has happened," and they all answered, "Yes, yes, as you truly say, two men wrestled and one ran into the garden," and so on. I don't understand it, "Something happened," and "Two men wrestled," don't seem to me at all the same; but I dare say I read old books of logic. Well, it's like that here. You seem to be all certain this man is a murderer. But I never said he was a murderer. I said he was the man we wanted. He is. I want him very much. I want him frightfully. I want him as the one thing we haven't got in the whole of this horrible case – a witness!'
>
> They all stared at him, but in a frowning fashion, like men trying to follow a sharp new turn of the argument. (Chesterton, 1929, p. 617)

It is odd that schools ignore training in perception of ambiguities, when so much of daily life involves sorting out meanings. I once was lost in the Fort Worth–Dallas airport because I interpreted a vertical arrow to mean that I needed to go up the stairs nearby; it was not until I realized that the arrow could also mean 'go further along this level in the way you are heading' that I could find what I was looking for. Alertness to the possibility of alternative interpretations makes a person more able to cope with many aspects of living, including learning. The series of pendulum lessons observed by Baird (1984) show how lack of this strategy in teacher and students causes confusion. The teacher had written on the board 'What factors affect period of the pendulum?'. We have already seen examples showing the students' vagueness about the meaning of period; it gradually became clear that they also interpreted 'factors' in ways that the teacher had not intended:

Observer: What do you have to do in this activity?

Student A: To find out what are all the outside factors which affect that period, that is factors other than the pendulum that affect it, like wind, bumping the desk . . . or the pendulum might change – well, the string might break . . .

Student M: . . . or it might slip, up the top, and take longer to make an oscillation.

Observer:	Do you see that, as part of the task, you can make any changes to the pendulum at all?
Student A:	Yes, I think so.
Observer:	What, for instance?
Student A:	Well, I suppose you could do anything to the pendulum ... like lengthening the string.
Observer:	Is that an outside factor?
Student A:	I suppose it isn't ...

(Baird, 1984, p. 311)

It took the teacher and the students several lessons to discover, with the observer's help, that they had meant different things by 'factor'. The probability that communications are open to alternative interpretations is not stressed in schools, so fewer people than might do acquire this useful strategy.

Planning

Planning involves setting out options, withholding action, and evaluating likely costs and rewards.

Almost everyone is able to set out options. We demonstrate this whenever we ask whether we will do this or that, or when we make any decision at all. Able people, though, apply the strategy more frequently and deliberately. They ask: Is there another way to do this? Which is better? How long will each take? I recall an army exercise in which soldiers were given half-a-dozen compass bearings and distances which they were to follow in a night march. All the soldiers but one followed leg after leg, struggling through creeks and across miles of rocky ground. The exception did a rough plot of the legs on the back of his hand and discovered that they led back to the starting point, so he moved off a short distance and rested until he heard the first of his weary fellows returning, when he rejoined them. The strategy of considering alternatives does not always lead to such dramatic savings in effort, but often makes learning easier. Unfortunately, schools do not teach this approach as well as they might.

Even if people are able to set out options, they may not apply that strategy because they are unable to withhold action and simply carry out the first thing that occurs to them. We call such people impulsive. They never get around to evaluating consequences. At the other extreme are the Hamlets of the world who hover between alternatives, unable to decide which course to follow.

Deciding between options involves judging the worth of rewards and the cost of penalties, estimating the likelihood of success, and selecting. Imagine a student at home on a weekend, revising science for an imminent examination. Options are to keep at the revision or to abandon it and go out with friends. Rewards in keeping on working are the sense of duty done and a greater chance of doing well in the examination, costs are the loss of time from enjoyment, withdrawal from social contact, a sense of drudgery. Other rewards and costs exist for going out. How the student values these rewards and costs depends on his or her goals and how clearly they have been established. Estimating the likelihood of success in carrying out either plan is not a major matter in this example, since presumably there is not much chance of failing to carry them out. The student could, though, think 'If I stay home I might be interrupted by my brother, so I might as well go out', or 'If I go out, I might not find any of my friends who will go with me, so I might as well work'. The student's choice involves balancing the judgements of rewards and penalties with the assessed likelihood of success for each option. The involvement of these strategies in personality is clear. Optimists are people who judge success probable: people with high external locus of control do not set out and consider options, because they believe the course of action will not be determined by themselves; anxious people give more weight to the penalties than the rewards in selecting.

Processing

Once the person has summed up the situation and decided what to do, new strategies come into play. In learning they include paraphrasing, which determines whether a communication is stored as propositions or left as a string, and associating, which determines whether the communication is treated as unrelated parts or as a coherent message.

It might be thought that paraphrasing is not a great problem, simply the sort of processing that almost everyone does without thinking. However, an experience I had in the course of some research makes me suspect that a noticeable proportion of secondary school students do not paraphrase. I was checking tenth grade students' memories for four sentences, which were shown to them on cards. The sessions were individual. In preliminary trials with graduate students I had found that some were unable to remember anything much from the cards after as short a time as two minutes, if they were distracted in that time by conversation or a task. Since the experiment required rather better performance than that, immediately after each card had been shown to the tenth grade students it was placed face down and they were asked, 'What was on the card?'. Most responded with a paraphrase of the information, but some gave the sentence verbatim. Their intonation and facial movements gave the impression that they were relying on a photographic recall. It appears a promising field for research to see whether such behaviour is common, whether it is consistent across oral and written communication, and whether it is related to poor recall of the information in the sentences but good recognition of their exact form. People who exhibit this behaviour consistently would be in difficulty in coping with learning, both in school and out. Training in paraphrasing could help them greatly.

Other processing strategies are operations on present knowledge retrieved from memory, rather than on communications received from someone else, as paraphrasing and associating tend (though not exclusively) to be. Such operations include generalizing, deducing and reflecting. All may lead to new elements of knowledge, particularly new propositions. Generalizing and deducing are common and clearly understood terms which need no description. By 'reflecting' I mean contemplation of a piece of knowledge, which consists of posing questions about it. Among these questions, those beginning with Why and How are important because they mark attempts to form explanations, which are central to understanding. Even if the question cannot be answered immediately, it prepares the person to receive the knowledge later. Someone who wonders why the active metal aluminium does not rust away is ready to absorb the proposition that aluminium oxide forms a layer impervious to oxygen. Another important set begins with What if; these questions are explorations of the causal relations between terms in propositions. A person who reflects on the statement 'The Earth is tilted at $23\frac{1}{2}^\circ$; this causes the seasons' and asks 'What if the angle were 0°, or 90°?' may generate new propositions and come to greater understanding of the statement. Schools train students to answer questions, but not to form them. That may explain the common observation that younger children ask more questions than older ones: our practices train it out of them.

It is possible to increase students' powers of reflection, and so make them more able. Brooks lists questions generated by his tenth grade students at the end of a two-week series of lessons on law, which reveal a good degree of reflection. Some examples:

What if a person broke both the civil and criminal laws – would we be charged for both?

Why are people brought up on many different charges at the same time (e.g. murder, assault, rape, etc.) when they could simply be tried for the most serious one?

If you can't be committed of a crime between the ages of 8–14, then what is there to stop this age group going on a crime rampage? If a group of 12-year-olds smashed the windows of every house and shop in Laverton, would it be assumed that they are under the influence of an adult, or would they simply be given a warning, or would they be put in a detention centre? N.B. The crime is an exaggeration.

If a wife kills her husband after 3 years of constant bashing from him, is she charged with murder, manslaughter or self defence?

Why are laws often given alternatives? e.g. someone commits a murder – put in jail for life and then gets released in eight or so years.

How would a person go about learning of the laws and their rights if not at school? Is there a place which records these matters and puts them on display for the public?

What if you were climbing across your fence and your neighbour hit you with the broom. You fell on him and he broke his neck. What would happen?

How would you convince the law that someone has infringed your rights if there was no witnesses?

(Brooks, 1986, p. 215)

Generation of reflective questions demonstrates a readiness to go beyond the imposed task. Learning is then being done for its own sake, not to satisfy the demands of the teacher. Baird and White (1982a) identified a spontaneous case of this in one adult learner, who contrasted sharply with others who stuck close to the information they were given and did not think about its relation to other knowledge they had. For these others the topic was complete in itself, and knowledge was a collection of discrete, unrelated packages. For the first learner, science knowledge fitted in with history, poetry and personal experiences. This fitting was not required by the task, but was something she did of her own accord. Going beyond the specific requirements is the essence of reflection, but where school science is treated as a closed rather than an open subject, reflection is not encouraged.

Other processing strategies are applied in monitoring the progress of a plan. When they are not operated, the activity may continue automatically: a common term for this is 'day-dreaming'. Monitoring is a part of consciousness. There is an anecdote about Newton which illustrates that even, or maybe particularly, the great sometimes cease to think about what they are doing: 'At some seldom times when he designed to dine in hall, would turn to the left hand and go out into the street, when making a stop he found his mistake, would hastily turn back, and then sometimes instead of going into hall, return to his chamber again' (Letter from Humphrey Newton to John Conduit, 1728; in Wintle and Kenin, 1978, p. 564).

In learning, monitoring involves comparing new knowledge, whether transmitted or self-generated, with what was known before: Did I know that before? Does it clash with what I believe? Did I predict it would turn out that way? How sensible is this? Does it seem correct? Associated with this strategy is another of assessment of one's understanding, and continuous checking of how well the learning is going: Do I understand what I am doing? How clear is this to me? Again unfortunately, it seems that few students are encouraged to develop these strategies. An experienced teacher reports a trick he played on tenth grade students that exposed their failure to check on the sense of what they were supposed to learn:

The turning point came on 27th February. After encouraging students to ask questions, think about what they are doing, show initiative etc. and getting passive or negative results, I wrote notes on the board which students mechanically copied down. I held a geography book in my hand and pretended to copy the two paragraphs shown below from the text.

Water
The degree of rainfall for each half year and the annual seasonal deficit are the systems which determine which areas will receive rain and which won't.

However, in planning where to plant crops it is not enough to know the system, one must also take account of the different levels within each seasonal system.

We must also know how much of the soil will be lost by evaporation.

Length of the daylight period
Plants depend on light. The daylight hours vary from town to town depending on altitudes. Towns in low lying areas depend largely on the degree of photosynthesis and rainfall – clouds create shade which affects people's vision.

The plant's visionary cycle and light condensation greatly affect the amount of hygration that can exist in a certain town at a certain time.

Melbourne's hygration can vary by 20 cm from Sydney's at any particular time.

This procedure was a spur of the moment decision. I made the nonsense notes up on the spot. My instructions were 'to copy the notes down from the board'. The topic under study was 'Agriculture', where students could be expected to use technical terms and definitions. I waited until all students had copied the notes and then asked if anyone had any questions – I asked this a number of times and, to my recollection, out of two Year 10 classes only one student per class had a question. One asked the meaning of a term used, the other hesitantly questioned whether soil could evaporate. I guess it was from this point on that I realised three things:

Firstly, I thought I had been teaching in a fashion that encouraged student involvement and initiative. I now realised that I had not been challenging the students enough. My reaction to these two classes was one of concern about my teaching methods.

Secondly, I was surprised to see to what extent students expect teachers to dictate and dominate class situations. Students either believe that teachers should not be questioned or believe that it is much easier not to get involved in class discussion.

Thirdly, that as a teacher I had an obligation to alter my teaching strategies. (Hynes, 1986, pp. 30–1)

The reaction of one of the students was:

In mid-term I, we were asked to copy two paragraphs of notes from the board so, being students, we did.

> We didn't think about what we were writing and, I didn't question it until I came to the term 'hygration' and the statement 'at any time Melbourne's hygration may differ from Sydney's by 20 cm'. Obviously, this prompted slight concern but I didn't query it, preferring to wait until I'd finished and listen to the explanation that generally followed note-taking. But the explanation didn't come. Instead we were instructed to ask questions about anything we didn't understand. A few of us did this, but not many. A total of about 5 questions were asked from a class of 27.
>
> Maybe we all understood, or believed we did. At least, we were all ready to leave the classroom then and there with our notes as they were, so imagine our embarrassment when we found the notes were meaningless, made up by Mr Hynes.
>
> Since then we have not trusted anything teachers have written on the board, questioning even the slightest phrase that doesn't make sense. (Dibley, 1986, p. 88)

I appreciate particularly the resignation of control implied by the words, 'so, being students, we did'.

A final, and important monitoring strategy in learning and other activities is checking whether the task has been completed: Have I finished? Do I really understand this well enough? What more would I have to do to understand it fully? If I don't understand fully, do I understand well enough to justify stopping? This can go further, into evaluating the quality of the learning beyond the imposed requirements: What use was all this? How will it affect what I do next? How should I try to remember it?

Teaching of cognitive strategies

It is one thing to sketch a number of cognitive strategies that influence the quality of learning, another to assert that schoolchildren can be taught them and trained in their use so that they will become more able. Is there any evidence that it can be done? The question is an important one, since the cognitive strategies view of abilities has the potential to bring about a revolution in teaching. Although at this time we should not expect too much, since the notion is new and has attracted only a small fraction of the effort that went into IQ testing and Piagetian studies, there should be some assurance that effort put into training in cognitive strategies will be rewarded.

Weinstein and Mayer (1986) provide an overview of the early work, much of which was concerned with comprehension in reading. They suggest that strategies should be grouped into elaborating, organizing, monitoring and affective responding types.

The techniques for encouraging elaboration strategies in the studies that Weinstein and Mayer review tend to be well-known teaching practices, such as requiring students to paraphrase and construct their own notes and summaries, to do problems (not merely repetitive exercises), and to respond to questions in class that involve reflection and construction of answers rather than simple recall. In addition to these widespread practices, Weinstein and Mayer mention the less common technique of having students create analogies. Analogies, or metaphors, are the theme of Gordon's method of synectics (1961). Originally devised to promote creativity in engineers, synectics can be used in secondary and at least the upper levels of primary schools to induce students to think about and extend facts. For instance, following a lesson on volcanoes, students could be asked to discuss how a volcano is like a tree. This can lead to revision of the information and to new thoughts about volcanoes: trees and volcanoes are generally cone-shaped, are roughly round in cross-section, have a part above and a part below the ground, enclose an upward movement of liquid which continues as long as they are alive, have dormant periods, grow, exude gases, spread solid material around, and so on. Flaws in the analogy would be made clear. Although there does not appear to have been any extensive research, one could expect that students who frequently participated in synectics exercises would come to create analogies spontaneously, and so would elaborate their knowledge habitually.

The organizational techniques that Weinstein and Mayer (1986) describe are, like synectics, systematic procedures that as yet are not used by teachers as a matter of course, although they could do much to make students better processors of text. The techniques are methods of determining the structure of a text, sorting out its principal themes and making clear the pattern and nature of the links within it from one concept or idea to another. Examples are the networking procedures of Holley and Dansereau (1984) and Meyer (1975; Meyer, Brandt and Bluth, 1980), which can be applied to science texts as well as those of other subjects. Cook (1982) devised a procedure specifically for science texts, in which students classify the form of a passage as an explanation or extension of a main principle, a listing of facts, a description of a sequence of steps, a sorting of objects or ideas into classes, or a comparison. These techniques are easy to teach, and various studies (e.g. Holley et al., 1979; Dansereau et al., 1983) confirm that their application results in more extensive processing of texts than the normal level, in which many students read sentences without discriminating between their importance and without thinking about the overall structure. We do not yet know, however, whether training in the techniques can become part of the normal program of schools, and whether it leads to permanent changes in the ways students learn. Like synectics, the techniques are promising but not established in practice.

One problem with techniques such as synectics and networking is that students can apply them with a minimum of reflection on what they are doing and what the purpose is. Their learning will then be unsatisfactory. Unfortunately, as Baird (1984) and Tasker (1981) found, students are often not aware of the purpose of a lesson, why they are learning some topic, or how well they are doing. They lack good comprehension monitoring strategies. Awareness of one's thinking and control of it is also called 'metacognition' (Brown, 1980; Flavell, 1976). Many experiments, mostly to do with reading passages of text, have been conducted to see whether students can be trained in metacognition (e.g. André and Anderson, 1978–9; Brown and Smiley, 1977; Ghatala et al., 1985; Markman, 1979; Paris and Jacobs, 1984; Wong and Jones, 1982). These have usually reported positive results.

Although the success of metacognition studies is encouraging, most of them involved relatively short-term interventions in an experimental setting, in which students had to learn some text that was not part of their normal program. It is more important, and more difficult and tedious, to find out whether metacognition and a permanent change in learning style can be promoted within a school program. Baird (1984, 1986) had some success in increasing metacognition when he worked for six months with one teacher and his ninth- and eleventh-grade science classes, although he found he had to struggle against ingrained habits of both the teacher and the students and their perceptions of what learning is about. The students had become used to a routine of passive acceptance of whatever the teacher presented, a routine that continued to be reinforced by practices in all their other subjects. Baird encouraged use of many of the self-questions that were listed earlier and was able to bring about more purposeful learning.

Metacognition principles were applied more widely in a continuation of Baird's work, a two-year action research study known as PEEL, the Project to Enhance Effective Learning. Teachers of science in a high school were joined by those of English, history, geography, commerce and legal studies in a concerted effort to promote metalearning, that is improved control by the students of their learning. The teachers' accounts of their experiences (Baird and Mitchell, 1986) describe many innovations that they used in helping their students become more able learners, such as assignments to write questions (not answers), classifying problems, concept maps, having the students write responses to assessments of their work, check lists of appropriate behaviours (Did I know what I was supposed to do? Did I ask questions?), letters between teacher and student in which students often pointed out things they did not understand, and interpretive discussions in which the teacher's views were withheld while the students had to state theirs and justify them. The teachers also identify difficulties with which they had to contend, and are frank about errors they made.

The success of the laboratory experiments and of Baird's programs in functioning schools in promoting metacognition encourage replacement of the psychometric and Piagetian level views of ability by the cognitive strategies view. The successes imply that schools could do more to improve learning by shifting the balance in their attention from almost total concern with content some way towards strategies. Adoption of the cognitive strategies view would involve another shift of social importance. Schools have an uneasy balance between two functions: they are producers, in that they are meant to take in students and release them years later in an improved, more knowledgeable state; but they are also sifters, a means of sorting out those who are fit to go on and qualify for the more prized professional occupations. The psychometric view fits well with the sifting role. The cognitive strategies view requires schools to turn away from sifting and to attend more to producing. This is not a comfortable idea. It places responsibilities on teachers and parents to foster the cognitive strategies and on students to acquire and practice them. It implies that revolutionary changes should occur in the organization of schools, in curricula and in teaching methods. The assignment of responsibilities and the revolutionary changes militate against acceptance of the cognitive strategies view, but if it should prevail it will bring about the major change to occur in the practice of teaching since the nineteenth century.

7

Attitudes

Both in and out of school, people say things such as 'If Tom had a better attitude to school he would learn better' and 'This new course should give the students a better attitude to biology' often enough to show that attitude is a common notion, understood as distinct from knowledge and ability and yet, like them, both a determinant and a consequence of learning. Psychologists see it in that way, too, and that is how I portrayed it in figure 2.1.

There is no substantial difference between the everyday and psychological uses of the word 'attitude', though in time one could develop if the psychological meaning becomes more precise. For the moment, as with many other constructs, the meaning of attitude remains vague and ambiguous. The ambiguity is exemplified by the range of techniques used to measure attitudes – the semantic differential (Osgood, Suci and Tannenbaum, 1957), Bogardus' intuitive interval scale (Bogardus, 1925), Thurstone's equal-appearing interval scale (Thurstone, 1928), Likert's summated ratings scale (Likert, 1932), observations of non-verbal physical reactions such as pupil dilation or skin conductivity, and others. The range of techniques does not in itself demonstrate that the concept of attitude is vague, for after all there are many methods for measuring the extremely precise concept of length. However, the measurements of a length correlate highly with each other, while those of attitudes do not (Ormerod and Wood, 1983). Dissatisfaction with the vagueness of the concept occasionally provokes attempts to do away with attitudes as a scientific term (e.g. Blumer, 1955; Doob, 1947), but so far it has survived all the criticisms, presumably because it is useful.

It is easier to experience or demonstrate an attitude than to define one. The term originally referred to a physical stance; we still use it that way in speaking of a 'prone attitude', for instance, and we can see it in Gilbert's song from HMS Pinafore about the requirements for a standard British sailor:

> His eyes should flash with an inborn fire,
> His brow with scorn be wrung;
> He never should bow down to a domineering frown,
> Or the tang of a tyrant tongue.
> His foot should stamp and his throat should growl,
> His hair should twirl and his face should scowl;
> His eyes should flash and his breast protrude,
> And this should be his customary attitude.
> (Gilbert, 1959, p. 79)

Confronted by someone fitting that description, we might well expect him to behave in a particular way. Common use often refers to potential behaviour, as in the phrase 'an aggressive attitude'. In a behaviourist psychology one might be content to have attitude defined in relation to behaviour, although even at the height (or depth) of behaviourism Allport was prepared to define attitude as 'a mental and neural state of readiness, organized through experience, exerting a directive or dynamic influence upon the individual's response to all objects and situations with which it is related' (Allport, 1935, p. 810). Later definitions include cognition: 'An attitude is a relatively enduring organization of beliefs around an object or situation predisposing one to respond in some preferential manner' (Rokeach, 1970, p. 112). We might add an affective component as well, emotions or the involuntary physical reactions associated with them, such as heart rate, production of adrenalin, concentration of blood at deeper levels of the body, laughter, tears.

Academics argue over the need to incorporate all three components – cognition, affect and tendency to behave – into the notion of attitude. Fishbein and Ajzen (1974) see affect as the determinant of the other two, and hence all that need be measured. For Bagozzi and Burnkrant (1979) attitude is the interplay of affect and cognition, with tendency to behave as a secondary consequence; while Kothandapani (1971) claims that all three components are equally involved in the nature of an attitude. While the nature of attitudes would be clarified if these positions were resolved, it is not crucial for that to happen for the present purpose of developing a psychology of the learning of science. I will take the position, substantially that of Bagozzi and Burnkrant, that an attitude to a concept such as science is the person's collection of beliefs about it, and episodes that are associated with it, that are linked with emotional reactions. The stimulation of these reactions affects decisions to engage in behaviour, such as choosing to take a science course, to read about scientific matters, or to adopt a science-related hobby.

The beliefs that are linked with emotions are generally propositions, such as 'dissection is disgusting', but can extend to strings and images. In chapter 3 I specified dimensions of propositions: the degree of social consensus and arbitrariness, and whether they are prescriptive, evaluative or descriptive. Propositions that are part of an attitude tend to be evaluative or prescriptive, and not universally acceded to: science is boring, people should not have to study science, boys are better than girls at science. Descriptive propositions are not excluded: there are many formulae in physics, people have to do lots of problems in physics. Descriptive propositions reflect and affect attitudes through secondary associations. If the person dislikes formulae and problems, then the examples represent a negative attitude to physics, but that is reversed if the person likes them. Thus the same descriptive propositions held by two people can stand for opposed attitudes, depending on the secondary associations.

People readily distinguish between their beliefs that are part of attitudes and those that are generally accepted and less arbitrary. The set of elements obtained in an interview, that was shown in figure 5.9, like all the other sets that were obtained in that study, contains only neutral, descriptive propositions. They were obtained by asking, What do you *know*? If the question had been, what do you *feel*, or *think*, or even *believe*, about the topic, presumably evaluative and prescriptive propositions would have been volunteered.

Formation of attitudes

Propositions can be acquired through experience or through social transmission. A child could study mathematics and find it difficult, and so form the proposition 'maths is hard'. Or he could be surrounded by peers who say 'maths is hard', and come to accept that. The peers themselves could have acquired the belief from others. Although there does not seem to be any evidence for this, one would expect attitudinal propositions, like the commitment to factual propositions, to be more stable when acquired through direct experience and also when they are supported by a high degree of social consensus. If a student finds that she enjoys working with chemicals, and is surrounded by others who say they enjoy it too, she is likely to acquire a belief that chemistry is fun, the basis for a stable attitude. It might take many subsequent negative experiences to shake that belief.

Although experience is a powerful determinant of belief, the strength of social transmission should not be underestimated. Groups build up social and political beliefs about other groups which they rarely if ever see, and may even be prepared to hazard their comfort and future by going to war because of them. Group-transmitted and -supported beliefs may transcend personal experience. In Nazi Germany there were pervasive beliefs about Jews. One Nazi leader observed that 'every German has his good Jew', referring to the direct experience non-Jews had of Jewish fellow-citizens; but despite that normally favourable experience the general negative attitude to Jews was held strongly and widely enough for abominable persecution to be accepted. We even have a saying, 'The exception proves the rule', to help us discount direct experience in favour of a socially transmitted stereotype.

Among school subjects, mathematics may have succumbed already to a social consensus of negative beliefs. Science is not yet so threatened. Studies commonly show that schoolchildren begin the study of science with favourable attitudes to it, though unfortunately many gradually come to regard it less positively (Campbell, 1972; Fraser, 1975; Gardner, 1974). This suggests that direct experience, not social transmission, is the cause of unfavourable attitudes to school science. While an indictment of the way science is taught, this is a relief because it means that something can be done about it by teachers without having to change the views of the whole of the community.

However, the community does have an influence. Outside school, beliefs about science are much more affected by social transmission than by direct experience. During the 1950s science got a good press, as new inventions and advances in engineering, coupled with a rising economy, brought unprecedented comforts within the reach of many people. Exciting achievements in space exploration captured imaginations. Later, science appeared more threatening. Pollution, desecration of the environment, the threat of annihilation in nuclear war, and medical tragedies like thalidomide, became part of the daily news and overshadowed those comforts in people's minds. Few were conscious of direct experience of these matters, but they were heard of so often that they became accepted as the fault of science; they were a threat caused by science even though scientists were prominent in identifying the threats and attempting to counter them. These socially induced beliefs are likely to affect attitudes to school science in time, if they are not already doing so.

Strings are important in the social transmission of attitudes. They are not inferred from experience, but are handed on and reinforced by repetition. The characters in *Animal Farm* (Orwell, 1945) learned the string 'Two legs good, four legs bad' by being told it and drilled in it, not by extracting it from their own experience. Although the full meaning of such a string may not be understood, it can mould a person's attitude: Orwell's animals responded positively to the pigs when they entered on two feet, and tried to copy them. Advertising agents, politicians and other propagandists have long been aware of the force of strings in determining attitudes: 'No taxation without representation', 'England needs you', 'A New Deal'. Although not meaningless in themselves, the repetitive use of strings in a social context almost divorces them from meaning. Why shouldn't there be taxation without representation, if needed services are provided? What does England need you for? A new deal of what? These probings of meaning are not essential for the string to influence, to be part of, an attitude.

Propagandists also appreciate the power of images in forming beliefs (see figure 7.1). Images are more important than strings in attitudes to science. Although I can recall many strings that have been used in politics and advertising, I cannot think of any that apply to science. On the other hand it is easy to find instances of images. The cartoon in figure 7.2 contains several of the features that Mead and Metraux (1957) identified in the standard image that American high school students had of scientists:

> *The scientist is a man who wears a white coat and works in a laboratory. He is elderly* or *middle aged and wears glasses. He is small,* sometimes *small and stout,* or *tall and thin. He* may be *bald. He* may *wear a beard,* may be *unshaven and unkempt. He* may be *stooped and tired.*
>
> *He is surrounded by equipment: test tubes, bunsen burners, flasks and bottles, a jungle gym of blown glass tubes and weird machines with dials. The sparkling white laboratory is full of sounds: the bubbling of liquids in test tubes and flasks, the squeaks and squeals of laboratory animals, the muttering voice of the scientist.*
>
> *He spends his days doing experiments. He pours chemicals from one test tube into another. He peers raptly through microscopes. He scans the heavens through a telescope [or a microscope!]. He experiments with plants and animals, cutting them apart, injecting serum into animals. He writes neatly in black notebooks.* (Mead and Metraux, 1957, pp. 386–7, italics in original)

The same features were evident in many of the 4,807 drawings that Chambers (1983) obtained from children in kindergarten to fifth grade in Canada, the United States and Australia. Among all those drawings there were only 28 women scientists, all drawn by girls.

The publicity given to Einstein late in his life, with his unusual hair and odd habits such as never wearing socks, both promoted and depended on the stereotype image. It is no wonder that female participation in science is much less than male. Other unflattering images are presented in stories and comic books, where the scientist is often portrayed as unworldly or even socially evil: consider Dr Jekyll and Mr Hyde, Frankenstein and Ian Fleming's Dr No.

Although images may most often be acquired through social transmission, they are also formed from experience, which may be a single dramatic event or a cumulation of similar ones. For instance, a succession of science teachers may lead to a composite image, in much the same way as recurring

Figure 7.1 Influencing attitudes through images (First World War poster by Norman Lindsay, used by permission of the Imperial War Museum)

Figure 7.2 Cartoon demonstrating the stereotypic image of scientists (from Hollowood, 1962)

episodes lead to scripts. A continued diet of drill exercises may lead to an image of science as slavery, or other experiences may give an image of a laboratory with coloured liquids, interesting smells and flickering lights. These images are virtually indistinguishable from generalized episodes.

We can see how a collection of beliefs builds up, forming an attitude to science, by imagining the experiences of a child. At first the infant has no propositions, strings, images, or episodes that relate to science, and so has no attitude to it. An early acquisition might be an image obtained from a picture book, or a proposition (which might initially be a string) picked out from a conversation between the parents: 'I see the French are going to explode another bomb in the Pacific. I hope their scientists go up with it.' The most likely source these days is television, where it is a matter of chance whether a positive or negative view of science is observed. Another source is the real world, where the child's experiences often are interpreted for her by adults: 'Put that worm down, Mary, It's dirty.' 'Look at the rainbow, Jean. Isn't it beautiful? Let's make our own rainbow with the hose.' The adult adds valuing words like 'dirty' and 'beautiful' to the child's experience, and they become labels on the episode. If enough natural phenomena receive negative labels, then an antipathy to science forms, while positive labels produce favourable attitudes. Absence of experiences and labels is unlikely, unless the child is cocooned. There will be experiences with toys. Some of these will give more pleasure than others, and some types of toy will be given to some types of child. Rich children will get more toys than poor. Boys rather than girls will be given construction kits and chemical sets. These differences in experience and the accompanying pleasure will lead to differences in the generalized episodes and the attached labels. A girl may turn to science because of envy of her brother's Meccano set, or away from it for the same reason. It depends on how the experience is interpreted.

It is hard to say how early in life differences in attitudes may form. Smail and Kelly (1984) observed striking differences at the end of elementary school between boys and girls in their liking for different strands of science. The boys were fond of physical science, the girls of natural history. The differences were firmly established, and it can be concluded that they had built up over a long period. It may well be that even before they get to elementary school children have acquired propositions and other memory elements that are the foundations of an attitude to science. One child has learned to enjoy fitting together hard-edged objects – boxes, blocks, nuts and bolts – and has learned that slugs are disgusting and gardens must be kept neat. Another had enjoyed playing with a cat and a dog, and has learned that gardens are for playing in and that plants should be left to grow. The children interpret the things their teacher says, and the experience the teacher arranges for them, in terms of these early experiences and beliefs, generally in such a way as to support the views already formed. Voluntary activities continue to drive apart the interests and attitudes of the two children. We not only do what we like, we come to like what we do, so the more time children spend in different activities the more their beliefs about them will diverge.

Eventually some of the activities of each child are given the label 'science'. If they have been enjoyable ones, then science is valued and fresh activities that come under the name of science are likely to be entered into. This smooth, incremental pattern may be interrupted dramatically. One unfortunate experience – a failure in an important examination, an injury in a science lesson, an unjustified reproof by a science teacher – can turn the child against science. On the other hand, an unexpected, pleasant experience may reveal joys and beauties that earlier experience closed the eyes to. The paradigm (which unfortunately deals with mathematics, not science, but is too good an illustration to waste) is the seventeenth-century philosopher Thomas Hobbes:

> He was 40 years old before he looked on Geometry; which happened accidentally. Being in a Gentleman's Library, Euclid's Elements lay open, and 'twas the 47 *El. libri* I. He read the Proposition. *By G*, sayd he (he would now and then sweare an emphaticall Oath by way of emphasis) *this is impossible*! So he reads the

Demonstration of it, which referred him back to such a Proposition; which proposition he read. That referred him back to another, which he also read. *Et sic deinceps* [and so on] that at last he was demonstratively convinced of that trueth. This made him in love with Geometry.

I have heard Mr Hobbes say that he was wont to draw lines on his thigh and on the sheetes, abed, and also multiply and divide. (Dick, 1949, p. 150)

These speculations on the formation of attitudes are not supported by any long-term longitudinal study. No one has followed from early childhood the accretion of beliefs that constitute an attitude to science. There have been many studies on the effect of courses of study on attitudes, but at best these cover only a year or two, usually in late secondary school (e.g. Fraser, 1980; Hasan and Billeh, 1975; Hofman, 1977; Kempa and Dubé, 1974; Krockover and Malcolm, 1978; Lazarowitz, Barufaldi and Huntsberger, 1978; Selmes, 1973; Sherwood and Herron, 1976; Tolman and Barufaldi, 1979).

According to my model of learning, to change an attitude involves adding new propositions, images, strings and episodes, and abandoning old ones. The old ones may stand in the way of the new. A belief that science is threatening will act against acquisition of another, that science is good. Recollection of negative episodes will hold one back from experiences that might lead to positive episodes.

Effect of attitudes on learning

So far we have been considering the ways in which attitudes are formed. The model also emphasizes that the cognitive and affective aspects of attitudes influence behaviour, which of course includes learning. However, it is not a simple relation since liking for an enterprise is not the sole determinant of engaging in it. Different contexts may activate different sets of beliefs or feelings, and can even influence physical states. A drab, dirty setting may let other feelings take over from interest in science and cause withdrawal from a scientific activity, or an attractive setting may encourage someone who normally has no warm feeling for science to take part. Goals come into it, too. Often, for long-term gains, people will do things that they find relatively distasteful. They may not like physics, for example, yet enrol for a course because it opens up career opportunities. Strangely, there has been little research on the effect attitudes have on performance. Most studies have been concerned with developing and comparing lists of attitudes, and on the effect curricula or methods of instruction have on attitudes (White and Tisher, 1986).

Perhaps the influence of attitudes on learning has been neglected because it seems self-evident and simple: if people are interested in science then they will learn it, and if they are not, they won't. But it is much more complicated than that, since for one thing attitudes encompass more than interest and extend to traits such as curiosity and appreciation, and for another the nature of a person's attitudes affects not only whether any learning occurs but also the style of that learning. In other words, attitudes influence the operation of cognitive strategies.

Imagine the uses three students make of the statement by their teacher that 'Banksias are indigenous to Australia'. One, who sees science solely as a means to an end, providing entry to attractive careers, learns it as a proposition to be recalled in an examination. As there is no value in it beyond the examination, there is no need to reflect on it, to relate it to other knowledge or to life outside the classroom. Another, who appreciates that science is a rich description of the natural world that adds to the quality of one's experiences, will recall episodes of seeing banksias in gardens and in forests, and of admiring their striking foliage and flowers. Perhaps images of May Gibbs' children's classic *Snugglepot and Cuddlepie* (Gibbs, 1946) will come to mind, a story in which the villains were 'bold, bad banksia-men' who pursued the gumblossom heroines. The 'banksia-men' were figures based on the tall seed cones of banksias, which can indeed be imagined as multi-mouthed creatures watching, watching as one walks through the Australian bush. None of that would be part of the first person's learning. It could be part of the third's, one who believes that science is a social enterprise, the construction over centuries of human endeavour of a potentially fallible body of knowledge in an attempt to bring order to our understanding of the universe. That person would also link the statement to knowledge of Joseph Banks and his voyage with Cook, and to the history of plant classification, neither of which would be of interest to the other two learners. Hence the three students' different attitudes to science lead to quite different processing of the statement. The quality of their learning is affected by their attitudes.

8

Perception of Context

It is remarkable how rarely the great learning theorists of the 1960s mentioned the effect of context upon learning or any other performance. Ausubel (1963, 1968), Gagné (1965), Bruner (1966) and Skinner (1968) describe learning almost as an abstract act, free of the effects of physical and social conditions that teachers know are so powerful. Perhaps the theorists saw context more as part of a theory of instruction than one of learning. Perhaps education then was more a part of psychology than a discipline in its own right. Much work in psychology consists of attempts to discover principles that apply universally, irrespective of conditions or persons; sometimes, as in Skinner's work, irrespective of species. In education, however, context cannot be ignored. To do so renders theories unworldly and inapplicable. The research that is based on those theories is planned and reported as if context does not matter, and consequently has little impact on practice. In practice, as Dunkin and Biddle pointed out (1974), context matters: it is one of the more important factors that determine what learning occurs.

Context is the conditions under which learning happens. It has many dimensions, and indeed it may have been its complexity that encouraged the earlier theorists to exclude it from their models. It has dimensions of place: in a school or outside, in a classroom or a laboratory. There are physical conditions of temperature and noise level. There is the population: how many people are present, their distribution of ages, how they are placed in the room. There are social dimensions, such as whether it is a formal, designated learning arrangement; whether the information that is presented will be required for an examination; whether attendance is voluntary or compelled. Many of the physical and social dimensions are obvious and easily assessed, but others of equal importance are subtle and subjective. There is the atmosphere of the situation, whether it is repressive or supportive; its structure, whether it is ordered or chaotic. There is the diversity of events, whether there are many or few, whether they are ephemeral and dynamic or enduring and static. There is where responsibility lies for management – who determines standards of behaviour, who initiates conversation, who decides what happens next. These, and other aspects of context, influence the learning that occurs.

Much depends on what the learner makes of the context. An observer may find the context threatening and repressive, the surroundings unpleasant and uncomfortable, but the learner may not perceive it that way at all. Photographs of old classrooms and accounts of what happened there may make them seem harsh and frightening places, but the children then may not have felt any worse about them than present-day pupils do about their schools.

Perception of context is closely bound up with the notion of scripts, the generalized episodes that were described in chapter 3. A script includes expectations for a situation, expectations about static physical conditions as well as for what will happen. To take the classic instance that Schank and Abelson (1977) gave for a script, a restaurant, we have different expectations for different sorts of places. We expect, and are more tolerant of, less luxurious conditions and less respectful manners in fast food takeaways than in expensive restaurants. It is the same with schools: people have become used to a certain standard of furnishing and to styles of behaviour that are not commonly found elsewhere.

Furnishing is not irrelevant to learning, for it carries a message. Museums, art galleries, theatres and cinemas all take seriously the effect physical conditions have on the quality of an experience. Galleries in particular attend to physical dimensions of rooms, lighting, carpeting, and the colour and texture of walls. Banks and corporations appreciate that conditions convey messages, and believe it is worth spending money on them. A meanly furnished school radiates the message that learning is not valued here, a message that dedicated teachers and administrators must then take extra effort to counter. Of course good education can occur in a log cabin or at the other end of a log from Rousseau, but in general better conditions make for better learning. Just because the Wright brothers worked in a bicycle shop and tried their products out on a beach there is no sense in Boeing trying to work that way.

A room in which there is a mix of interesting permanent and temporary displays teaches much good science. An untidy room, or a bare one, or one with dull, ancient posters and exhibits that have been there so long that no one sees them any more, teaches bad things. Science teachers should, therefore, attend to the messages that their classrooms convey. They are better placed to do that than teachers of most other disciplines, firstly because they are more likely to have a special room assigned to them for most of their lessons, while mathematics and language teachers and others often have to work in a succession of communal rooms; and secondly because there is such a range of fascinating possibilities – the whole universe can be displayed as the subject.

Though furnishings affect learning, behaviour is more important. And just as the script for school furnishings lags behind those for most other institutions, so the scripts for behaviour in schools have touches of the nineteenth century. Teachers act and speak in schools in ways that would look odd and unacceptable in other situations, and, despite recent rumblings, students acquiesce to authority in a way that has become less common elsewhere.

Classroom contexts are measured against a small number of stable scripts. If what is happening fits one of these scripts, everyone will know how to behave and will feel comfortable. If it does not fit, people will not know what to do and will behave unpredictably until the situation is resolved and incorporated in a script. This has important implications for the introduction of new teaching methods or attempts to change the ways in which students learn.

Each school has a general script, and each of its subgroups evolves its own specialized set of scripts within the general one. Students and teachers learn how to behave in each script, so that certain things can be done or said in the playground that are impossible in the classroom. Different classes have different scripts, though in most schools there are only minor variations except across grade levels, where older students would be outraged if they were expected to conform to the lower grade script and the younger ones would be puzzled to know what is allowed if they were placed in the seniors' script.

Scripts determine what the teacher tries to teach, and the manner of teaching, and what and how the students try to learn. More general aspects of context determine the range of scripts that may be developed. For instance, an individual teacher and class cannot easily escape from the social expectation that the teacher will maintain order and the students will be quiet and submissive, even if together they believe that they have discovered a more effective way of working. Teachers who believe that they have no ethical right to enforce compliance with their wishes cannot implement that belief in a normal school because the students would not then know how to behave. The ensuing chaos would make the situation intolerable. A student who thinks that good behaviour consists of questioning why the lesson is necessary and what everything means, in order to understand and learn better, is also outside the usual range of scripts and will soon find that such behaviour is not permissible.

An important example of how perception of context affects learning is the general acceptance that secondary schools are concerned with separate subject disciplines. That script determines how students form and search out associations between elements of knowledge. Students tend to think this lesson is part of biology, so possible links with physics, chemistry, mathematics, history, civics or music are not contemplated. Attempts by one teacher to get students to look for associations between one body of knowledge and another will have to be long-maintained and frequent if the general script is to be overcome. Science will often be seen by students as a circumscribed body of knowledge, which will make it hard for a teacher to present it as an integral part of social studies or history: he may be met by a chorus of objections, 'Sir, sir, that isn't science, it's history'. Scripts in secondary schools are especially difficult to change, because the students see several teachers a day at short intervals. Teachers in elementary schools, who in most countries have a whole year with the one group of children, are better placed to introduce their own scripts.

Each classroom group develops sub-scripts for varieties of occasion. Even defiance of the teacher, rejection of the major script, has its accepted rules. Only a limited range of behaviours is contemplated, not, in most cases, extending to pretending to be crucified to the wall or undressing. The teacher's reactions to defiance also follow a script. In some places a cuff over the head is acceptable, in others not. In some countries corporal punishment is part of the script, and then it is expected to be used. An illustration is provided by Mercurio (1972), an American who studied the use of caning in New Zealand two decades ago. Caning was not part of the scripts Mercurio was used to, but to his surprise the New Zealand students and teachers accepted caning as a normal and reasonable part of school life. Teachers opposed to it were still expected to cane offenders, and, as I found myself when I taught in New Zealand for a year, had a harder time if they refused to conform to the script.

Scripts exist for how the students and teachers are to behave when the student does not understand the information being transmitted. In few schools are students expected to hold up the lesson until the point is clear to them. At best they are expected to wait until a time that they and the teacher recognize as convenient, and ask for help then. The teacher is expected to provide a few minutes only of personal attention; the script does not extend to long-term tutoring. This script has a crucial effect on the quality of learning, because it means that students have to be satisfied with less understanding than they could attain if they had more control over communication with the teacher.

The placement of control in a script affects learning. The main script in most secondary school classes has the teacher out in front doing most of the talking – about two-thirds of it according to Flanders (1970) – and asking a lot of questions. The answers to the questions are not to inform the teacher, for it is recognized that the teacher knows them already. The script has the teacher so much in control that the students may not get up or talk without permission. Their role is to respond to the demands of the teacher. They are not there to initiate activities of any kind. This script is astonishingly pervasive, applying in school systems in many countries. For many subjects it is the only script most teachers have.

Science classes usually have at least two major scripts – one for teacher–talk sessions and one for the laboratory. Some teachers and students develop several scripts, say two or three alternatives for the classroom plus one for the laboratory and another for excursions. Each script constitutes a different context, and each affects the sorts of things that are learned. Teacher domination of talk, for instance, tends to disseminate the message that science is facts, a collection of right answers determined by authority. A different script, say of discussion and argument, presents science as a set of opinions constructed from and supported by personal observations.

The introduction of a new script is not a simple matter. As scripts are generalized episodes there has to be time for a number of experiences to occur and be absorbed. People have to learn how to learn from a discussion or an excursion. The first few times young students go on a science excursion they are too busy acquiring the script of how to behave overtly in this new experience to alter their procedures of learning. Their procedure fits the context of a teacher–talk session, and does not work in the new context. They learn little. In time, with more excursions and guidance of what to do, they may develop new, effective ways of learning. Only too often, however, they do not.

To add a new script for a new context, such as the science laboratory or the excursion, is time-consuming and needs care. It is, however, easy compared with the task of changing perception of a familiar context, which is necessary when we try to change the style and quality of learning that occurs in classrooms. If the students' perception of the science classroom is a place where authority delivers facts that are important for examinations but irrelevant to life outside school, they will learn certain things and reject others. If we want them to see science as a means of enriching personal life through making surroundings more interesting and comprehensible, then we have to change their perception of the context of the classroom. The same applies to teaching. To change the way people teach requires a change in their perception of the context in which they work.

There are two ways of going about this. One is to change the context itself; the other is to tackle the perception. Neither is easy. The context is determined by the mix of social forces, outlined in chapter 1, in a system of checks and balances so stable that it is hard for an individual science teacher to feel that much can be done about it. However, it is not impossible. Although the inertia of whole school systems is immense, within their own classrooms teachers can begin to shift the context by introducing new scripts bit by bit. They can reward initiative, giving credit for good questions by students as well as for good answers. They can introduce discussions, on topics like 'Why are trees in cold and tropical climates evergreen but those in temperate climates deciduous?', in which they withhold their own views and so train the students to construct and defend their own opinions. They can try different teaching procedures, such as Suchman's inquiry training (1966) or Gordon's synectics (1961), to broaden their students' skills of learning. They can use new methods of assessment, such as the concept maps and prediction–observation–explanation techniques described in chapter 5. Those introductions would change students' perceptions of the purpose and procedures of learning. Context is a powerful determinant of learning, and as long as it remains shabby and limited, so learning will be mean and limited. We need to attend to the context in which science is learned, and to how it is perceived, if we are to improve the quality of learning.

9

Learning

A model of learning

Chapter 2 presented a model of the causal relations between constructs that determine learning, constructs that were described in detail in chapters 3 to 8. That model is now to be supported with another – a model that provides an account of what happens when someone learns science. Its sources are information processing and constructivist theories, which, though different, complement rather than conflict with each other.

Since distinctions between subjects are artificial and often blurred, many features of the account apply to knowledge beyond science. Science does, however, have special characteristics that make some issues more, and others less, prominent than they would be in a general description of learning. Especially important is the way in which the information presented by teachers and texts is filtered through a system of beliefs that the learner has established from observations of the world, a process that can lead to rejection or amendment of the information or to a change in beliefs.

The model summarized in figure 9.1 is typical of information-processing accounts, portraying the learner as attending to a selection of stimuli from a multitude of surrounding events, translating the physical stimuli into meaningful mental forms, then processing the forms to a greater or lesser extent so that they are more or less integrated with older knowledge. The act of processing is equivalent to the construction of meaning.

The figure does not represent adequately the complexity of learning, which is interactive rather than simply sequential, with what is already known affecting what is selected and the translation that is made of it. Neither, really, can an extended written description do so. The actuality of learning is more subtle than a model can show. However, even if the model

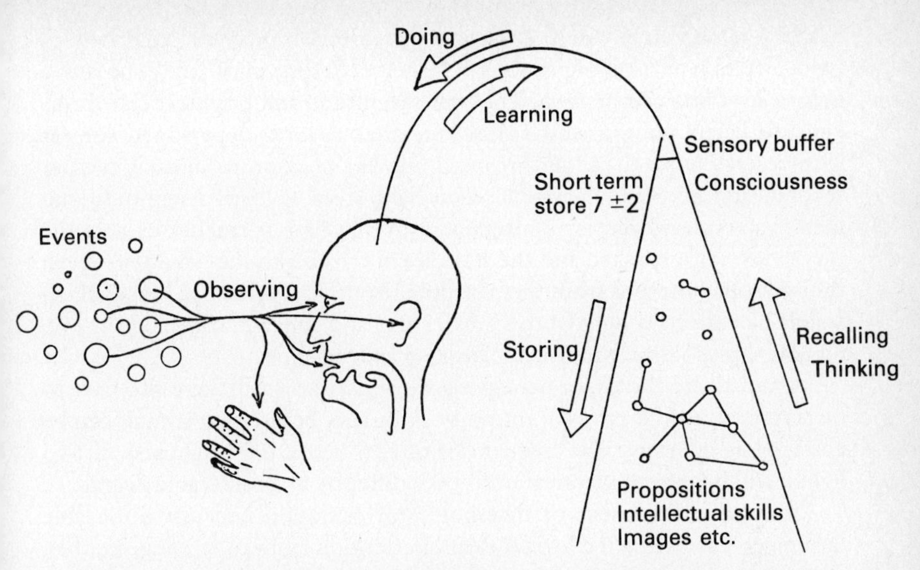

Figure 9.1 Representation of an information-processing model of learning

is only a shadow of reality, it serves a purpose if it makes the process less mysterious and more comprehensible, and indicates how learning may be fostered.

Selection and translation of events

The start of the sequence has the learner surrounded by events. The learner's body contains specialized receptors which are sensitive to four sorts of physical consequences of events, which are experienced as five senses. There is electromagnetic radiation, some frequencies of which are experienced as sight and some other frequencies as heat, while many other frequencies are not detected. There are variations in air pressure, of which a certain range of frequencies is experienced as sound. There is physical pressure on the skin, which is experienced as touch, and there is absorption of molecules by mucuous membranes in the mouth and nose, which is experienced as taste and smell.

Each of these stimuli has to be above a certain threshold intensity before the nervous system is triggered and a sensation is experienced. One stray molecule of ammonia is not enough for the person to smell the gas. Normally we do not smell metals because their vapour pressure is so low that insufficient atoms from them enter our nostrils, but when the metal is heated the vapour pressure rises and the threshold may be passed. With light, a certain number of photons per interval of time must strike the retina before anything can be seen. The first magnitude and brightest star Sirius emits sufficient photons and is close enough to us for each person looking at it to receive more than the threshold number of photons in each eye per second, and so Sirius is visible. The light from seventh magnitude and fainter stars, however, is so attenuated by the time it reaches us that the threshold is not reached and the stars are not visible to the naked eye, even though some photons from them strike the retina. A telescope makes them visible by collecting light from a greater area and concentrating it at the eye. Similar thresholds exist for taste, hearing and touch.

As well as thresholds for perceiving any sensation at all, there also have to be certain size differences in intensity or quality before two stimuli can be detected as different. The frequencies of two notes, the brightness of two lights, will be judged identical until they differ by a considerable degree.

Although these issues of thresholds for sensation and just noticeable difference have attracted a great deal of attention from psychologists, they are of less interest in education, where a more crucial phenomenon is that we are not conscious of all the stimuli that are above their thresholds. Although strong enough to be experienced, many stimuli are filtered out from consciousness and ignored. It is a remarkable feature of higher animals that attention is directable. As I write this I have my feet on a rail under my desk. Until now I have not been conscious of the stimuli my feet were receiving and the message they were transmitting to my brain. Now, in searching for an example, I have thought of my feet and can concentrate on their message. As it is not an important message, I will probably tune it out again soon. This power of direction of attention is vital to our functioning, and of importance in education. It is a major feature in determining what is learned.

The teacher intends the pupils to focus on certain stimuli from their surroundings. It is unfortunate for the teacher's plans when the pupils attend to something else. They are in the classroom, with its various sights, sounds, smells and physical feelings: birds are singing outside and light is streaming through the windows, the teacher is talking and writing on the board, other students are doing things, there are colourful posters on the wall, the principal is walking down the corridor. Out of all this the student has to focus on one event at a time. That is the other important feature of consciousness besides thresholds: we are serial processors, in the sense that we attend to one thing at a time. We are capable of rapid alternations of attention, which give some illusion of being involved in more than one situation, but really we are not. It is easy to demonstrate this: have two people simultaneously say different things, then try to repeat what each said. Either all of one message will be recalled, or bits of both, but not all of both. Listening to music while reading or conversing may appear to be an exception, but in fact all that is happening is an occasional momentary check that the music is still going on while the bulk of attention is given to the other activity. Or alternatively, some of us have well-developed social skills of appearing to carry on conversations while really attending to a book or music.

Selection of events for attention is obviously vital in learning. The principles of selection are summed up in the equation that the probability of selection of an event is a function of attributes of the event, attributes of the observer and the interaction of the two sets of attributes.

Attributes of events

The attributes of events which tend to make them noticed are properties that can be determined more or less objectively. The greater the energy involved in the stimulus – the more intense the light, the greater the amplitude of the sound wave, the greater the physical pressure, the greater the concentration of chemicals absorbed – the more likely the event is to be noticed. Other, rather more subtle, properties are important too. The eye sees colour and shape distinctly in only a small solid angle of a few degrees around the direction it is looking. The very notion of looking in a particular direction recognizes that we see clearly only in a small angle. To experience this narrow cone of vision, hold the eye on a point while someone brings an object slowly round from the side of your head to the front, at about 50 cm distance. Unless your eye has flicked to the side, you will not be able to tell the object's colour and outline shape until it is within about 20° from the 'line of vision'; and if it has fine detail such as writing on it you will not be able to pick that out until it is within a degree or two of the centre line. Although we cannot determine much detail on the periphery of vision, we are sensitive there to motion. Something moving fast on the edge of sight attracts attention. There are obvious evolutionary advantages in this: someone who ignores fast-moving objects is less likely to survive to pass on attributes to progeny.

As well as motion and absolute intensity of a stimulus, the relative strength of a stimulus compared with the rest of the field affects what is noticed. In 1983 a helicopter was flown in daytime into the sea off the Scilly Isles: although the sea was reflecting a lot of light, there was not sufficient contrast between it and the sky for the pilot to be aware of its nearness. Relative strength matters negatively as well as positively. If the rest of the field is at a uniform level of high intensity then a part that is at a low intensity is noticeable. Astronomers are aware of this: little or no light comes to us from dark gas clouds such as the Coalsack nebula, yet they stand out against a background of bright illumination from the Milky Way. The very words on this page are noticeable because little light reaches the eye from their black surfaces compared with that from the surrounding white paper. Contrast is an attribute that matters. Contrast can be in an instantaneous field or in a sequence of events. Teachers are aware of this: they may be talking at a steady tone or loudness, when to emphasize a point they raise or lower markedly the intensity or the pitch of the sound. They also know that an event which contrasts sharply with what has gone before is more likely to be attended to.

Attributes of the observer

Attributes of the observer also affect the selection of events. Among them is the general level of alertness. When this is very low, as in sleep, a stimulus must reach a very high intensity to be noticed. Sleep is an extreme level, but even in the wakeful state there are variations in alertness. These variations may have a physical cause, such as tiredness or illness, or the effect of a stimulant or a depressive drug. They may also have an emotional or a cognitive cause. Absorption in another activity may prevent new stimuli from being noticed, or a judgement may have been made about the general importance of what is going on.

Level of alertness is controllable. We can decide to be alert or not, within the limits of our physical state (if drugged or drowsy the limit is lowered). The limits can be expanded by training: in *The Psychology of Consciousness*, Ornstein (1972) describes some practical techniques for enhancing perception. Derived in most cases from Eastern practices such as yoga, these techniques involve a period of resistance to the continuous stimulation received from external distractions or from one's own thoughts about the cares of the day. One procedure is to concentrate on a koan or mantra, a saying of no or obscure meaning. These techniques, even when practised at a beginner's level, are effective in raising alertness to external stimuli, perhaps too effective since for a period afterwards even the most ordinary situation produces floods of perceptions such as the texture of the paving beneath one's shoes, the play of light on a wall, the cracks in bricks, the touch of a breeze on one's hands and distant sounds. It is tiring to be conscious of so much: we need to filter out most of what happens around us in order to function without exhaustion. However, it is not altogether silly to think of schools training students in alertness. Elementary schools already give some attention to alertness, by having rest and sleep periods in the junior grades. Upper primary, secondary and tertiary students are expected to maintain a steady, high level of alertness throughout the day, which may not be realistic. Thoughtful teachers are aware of this, and alter activities to suit their students' states. They hold off beginning a difficult topic when they perceive that the students are not alert, instead giving them some routine work to round off the previous topic.

Another attribute that affects selection is the range of cognitive strategies available to the observer. An observer who has developed the strategy of sorting out relevant from irrelevant is likely to make a different selection from one who treats all stimuli as of equal value. An observer who has the strategy of attending to the task in hand will select different stimuli from an observer who is easily distracted. One who has the strategy of searching out the meaning behind an event will focus on different things from one who treats all events as mere occurrences of no special significance.

Interaction between events and observer

At least as important as the direct influences of the attributes of the events and those of the observer are the interactions between them. People focus on the aspects of their surroundings that are relevant to their current purposes. Someone engaged in reading a book can filter out quite energetic stimuli such as the sound of dishes being washed in the next room, and be genuinely unaware that such activity is going on. A student in the laboratory can be so engrossed in an activity that he does not hear an instruction from the teacher. Absorption can, however, be broken into, by a light touch or by an even more energetic stimulus of the type that is being ignored. This implies that the environment is being scanned, and events judged and discarded unconsciously. An illustration of this scanning is provided by one's sensitivity to one's own name. At a party, talking with two or three others, you may be unaware of the conversations of other groups beyond an occasional flicker of attention that checks they are present, but if your name is spoken in one of those conversations you are likely to be instantly conscious of it.

Selection is also affected by whether the observer finds the event unusual, interesting or understandable. Before the event can be seen as any of these it has to be interpreted, which involves constructing meaning from the stimuli that are received, usually by grouping them into patterns that correspond to objects that have been experienced before. Thus figure 9.2 is not perceived as firings of a large number of individual neurones at the back of the eye, but as an arrangement of shapes.

Construction of patterns is such a fundamental human characteristic that we tend to forget that the patterns are learned. Learning of patterns is most obvious in babies. Although a baby's eyes register light from the moment of

Figure 9.2 Arrangement of shapes which people can see as an object (from Street, 1931)

birth, she does not 'see' in the sense of having pictures in the mind. Seeing is an interpretation of retinal stimulation, and is learned. Gradually, as the baby handles objects and becomes used to regularities in retinal stimulation through looking at the same things many times, she comes to see the world as a collection of organized patterns instead of an undifferentiated continuum of light. This implies that a child, or other higher organism, which was subjected from birth to a continuous random and non-recurring sequence of visual experiences would not learn to see.

Our drive to construct patterns is very strong. We do it even when the pattern is not really present, as in the illusory contrasts shown in figure 9.3. Although the figure does not contain a shape with three curved sides or a five-pointed star, most people construct them. In other instances alternative patterns may be constructed, so that the world is seen in a quite different way, as in figure 9.4 which can be seen as either a young woman or an old one.

Nearly all of the learning of patterns occurs in early life, and is not an issue in formal schooling. There are, however, many occasions in the learning of science when it matters a lot whether particular patterns have been acquired or not. A classic instance is learning to see with a microscope, which was immoralized by James Thurber (1933) in *My Life and Hard Times*.

Figure 9.3 Illusory contrasts: the black shapes influence observers to construct white shapes between them

Figure 9.4 Drawing that an observer can construct as either an old or a young woman

University Days

I passed all the other courses that I took at my University, but I could never pass botany. This was because all botany students had to spend several hours a week in a laboratory looking through a microscope at plant cells, and I could never see through a microscope. I never once saw a cell through a microscope. This used to enrage my instructor. He would wander around the laboratory pleased with the progress all the students were making in drawing the involved and, so I am told, interesting structure of flower cells, until he came to me. I would just be standing there, 'I can't see anything,' I would say. He would begin patiently enough, explaining how anybody can see through a microscope, but he would always end up in a fury, claiming that I could *too* see through a microscope but just pretended that I couldn't. 'It takes away from the beauty of flowers anyway,' I used to tell him. 'We are not concerned with beauty in this course,' he would say. 'We are concerned solely with what I may call the *mechanics* of flars.' 'Well,' I'd say, 'I can't see anything.' 'Try it just once again,' he'd say, and I would put my eye to the microscope and see nothing at all, except now and again a nebulous milky substance – a phenomenon of maladjustment. You were supposed to see a vivid, restless clockwork of sharply defined plant cells. 'I see what looks like a lot of milk,' I would tell him. This, he claimed, was the result of my not having adjusted the microscope properly, so he would readjust it for me, or rather, for himself. And I would look again and see milk.

I finally took a deferred pass, as they called it, and waited a year and tried again. (You had to pass one of the biological sciences or you couldn't graduate.) The professor had come back from vacation brown as a berry, bright-eyed, and eager to explain cell-structure again to his classes. 'Well,' he said to me, cheerily, when we met in the first laboratory hour of the semester, 'we're going to see cells this time, aren't we?' 'Yes, sir,' I said. Students to right of me and to left of me and in front of me were seeing cells; what's more, they were quietly drawing pictures of them in their notebooks. Of course, I didn't see anything.

'We'll try it,' the professor said to me, grimly, 'with every adjustment of the microscope known to man. As God is my witness, I'll arrange this glass so that you see cells through it or I'll give up teaching. In twenty-two years of botany, I –' He cut off abruptly for he was beginning to quiver all over, like Lionel Barrymore, and he genuinely wished to hold onto his temper; his scenes with me had taken a great deal out of him.

So we tried it with every adjustment of the microscope known to man. With only one of them did I see anything but blackness or the familiar lacteal opacity, and that time I saw, to my pleasure and amazement, a variegated constellation of flecks, specks, and dots. These I hastily drew. The instructor, noting my activity, came back from an adjoining desk, a smile on his lips and his eyebrows high in hope. He looked at my cell drawing. 'What's that?' he demanded, with a hint of a squeal in his voice. 'That's what I saw,' I said. 'You didn't, you didn't, you *did*n't!' he screamed, losing control of his temper instantly, and he bent over and squinted into the microscope. His head snapped up. 'That's your eye!' he shouted. 'You've fixed the lens so that it reflects! You've drawn your eye!' (From *My Life and Hard Times*. Copyright © James Thurber, 1933, 1961; published by Harper and Row Inc. Reproduced with the permission of Rosemary Thurber.)

Since patterns are learned, and alternative patterns may be constructed from a set of stimuli, seeing, and use of the other senses, involve imposing meaning. The patterns we construct are interpretable.

Figures 9.5 and 9.6 are usually seen as objects rather than collections of lines. We tend to complete lines from two collinear segments, so figure 9.5 is interpreted as a drawing of a shed. We operate rules of placement, so that figure 9.6 is seen as a three-dimensional set of blocks. Almost everyone will have assumed that the long horizontal lines represent a single block, although it could just as well be two, and that the triangle 'behind' it continues to the surface on which the blocks stand.

Since we tend to see events as collections of meaningful, or at least interpretable, units, selection is bound up with translation. If we cannot combine a set of stimuli into a unit, we are not likely to select them for observation. The simple sequential representation of figure 9.1 implies that sensations are held in the sensory buffer and some are selected for attention in the short-term memory. As selection has to follow translation, it is more complicated than that. Translation involves long-term memory, since we have to know about sheds and blocks or similar objects in order to see them.

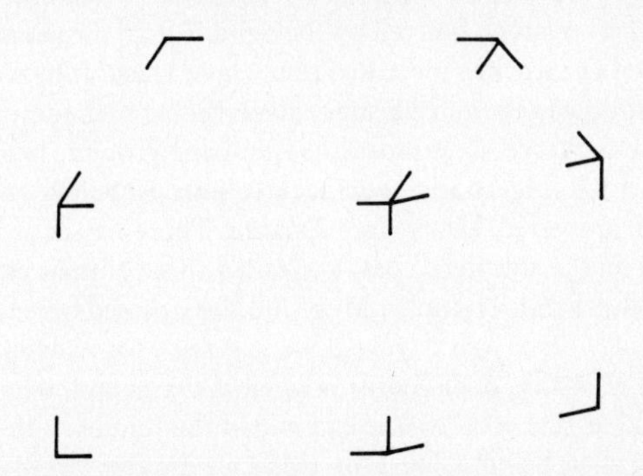

Figure 9.5 Eight marks that most people see as a shed

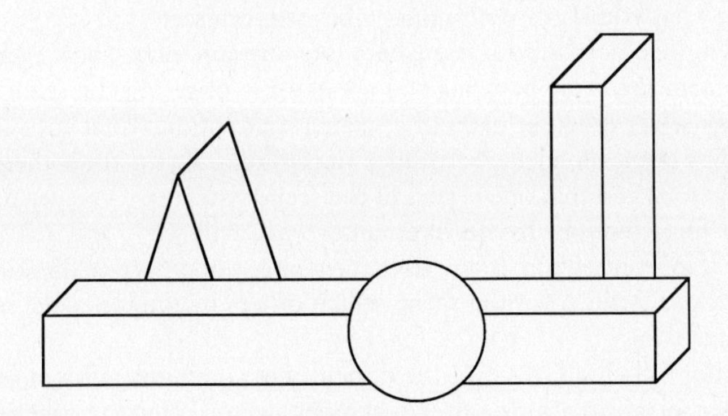

Figure 9.6 Two-dimensional drawing that people construct as three-dimensional blocks

The act of translation is automatic and not subject to much control at all. When I look out the window, I see recognizable objects immediately and apparently without effort. Only rarely is there a pause while the scene clears, so to speak. Those rare occasions, which usually arise through initial difficulty in judging the distance of an object, are revealing about the fact that translation has to occur. Usually, however, it is not a conscious process.

Knowledge affects sensation, so teacher and student may organize the stimuli they receive in different ways, and so may experience different sensations even when in the same surroundings. They may actually see different things when looking from the same vantage point in the same direction. Thus the student may see no more than an undifferentiated field of grass, while the geologist, or the botanist, sees it as a number of highly contrasted areas, sharply defined. The difference between what skilled and unskilled observers see can be dramatic:

A few days later we passed some tracks. I was not even certain that they were made by camels, for they were much blurred by the wind. Sultan turned to a grey-bearded man who was noted as a tracker and asked him whose tracks they were, and the man turned aside and followed them for a short distance. He then jumped off his camel, looked at the tracks where they crossed some hard ground, broke some camel-droppings between his fingers and rode back to join us. Sultan asked, 'Who were they?' and the man answered, 'They were Awamir. There are six of them. They have raided the Junuba on the southern coast and taken three of their camels. They have come here from Sahma and watered at Mugshin. They passed here ten days ago.' We had seen no Arabs for seventeen days and we saw none for a further twenty-seven. On our return we met some Bait Kathir near Jabal Qarra and, when we exchanged our news, they told us that six Awamir had raided the Junuba, killed three of them, and taken three of their camels. The only thing we did not already know was that they had killed anyone. (Thesiger, 1959, pp. 51–2)

Clearly, what you know determines what you can see.

Even more crucial in education, because more common, is that even if two people form the same patterns and are capable of seeing the same things, they may select different parts of a scene for attention. What they select is affected by their knowledge, attitudes and abilities. If you know that snakes are dangerous and that one is part of the scene you are in, you are likely to select it for attention. If you like money, you will see it and attend to it when it is in front of you. If you have the strategy of reflective thinking, you will pick out items from the scene which others might pass over as of no importance.

Selection is behind much of the difficulty in communication in teaching and other social situations. Often when one person is talking others attend only intermittently, preferring to select from a book or to hold off all external stimuli while carrying on with their own thinking. Either what is being said is ignored entirely, or gaps in the reception are filled in as much as possible with inferences or guesses about what the speaker might have said. The more routine the situation, the more likely we are to behave like that. Teachers, of course, are rarely prepared to tolerate lack of concentration on the events they want their students to select, and many of their skills are to do with maintaining attention.

Short-term memory

According to the model, when events have been selected and translated they are held in a short-term memory store. Selection of external events is not the only source of elements for the short-term memory. They can also be recalled from long-term memory, through perception of a link with an element already present. Thus if the notion of 'wall' enters the short-term store through external perception, this may call up a string: 'O sweet and lovely wall, Show me thy chink, to blink through with mine eyne!' and that in turn an episode of taking part in or seeing *A Midsummer Night's Dream*, and perhaps an image of a cell wall can be triggered, followed by propositions about plant and animal cells. A rapid succession of elements can flick through the short-term store.

The properties of the short-term have been subjected to intense study, almost entirely by manipulation of external stimuli. The short-term store is found to have a very limited capacity, and items can be held in it for only a brief duration unless they are rehearsed. One of the best-known results of research on memory is Miller's demonstration that most people can hold only about 7(±2) units in the short-term store (1956). If someone reads to you a sequence of numbers, you can repeat them if there are seven digits or fewer, but longer sequences can rarely be repeated without error. Less well known is Broadbent's revision of Miller's conclusion, suggesting that the real number is rather less, three rather than seven (1975). An experience of the lower estimate can be obtained through the following exercise. Imagine a sphere, a cube and a pyramid side by side. Make mental manipulations of them, such as placing one above the other or standing one on the other two. Now add a rod. It is much harder to visualize all four objects than it was for three. Try adding further objects such as a half-moon and a star, and you will find that they cannot all be kept in the picture at the same time. You can keep attending to them all by shifting to and fro across the arrangement of objects, but while picturing one group of about three or four the others fade.

Whether it is seven things or three that can be thought of at a time, the reality is more complicated than a static model with a number of pigeon-holes to be filled can represent. It is more as if, rather than a distinct short-term store, there is a depth of consciousness, with things at the surface available for inspection and other things retrievable only if there are lines to them which can be used to pull them up. A series of numbers has no specific lines, and once lost from the surface is gone for ever. Certainly it is clear that you are aware of more than the seven numbers when they are being read: you are aware of the nature of the task, so are remembering that; you are aware that someone has entered the room, that a door is banging, and so on. It seems that you are holding more than seven things in mind at once, even if not focusing directly on more than two or three.

Even though models are not good at representing the complex reality of memory and consciousness, the notion of a short-term store which can hold only a few things at a time is a useful one in learning, for it makes us think about what we mean by 'things'. We chunk the world, that is combine our sensations into a small number of patterns, as was illustrated by figures 9.5 and 9.6. When the string of numbers is read out, we group the sounds into the separate numbers. We hear the reader say 'seven' as a unit, a single chunk, not as two separate syllables or an even greater number of distinct sounds. Chunking is a function of knowledge: we know 'seven', otherwise we would not hear it as a unit. Foreign speech is hard to interpret even if one is moderately facile in reading it, until the pattern of chunks is established. One aspect of learning is to see the world in fewer, larger chunks. It is easy to illustrate this. Look at figure 9.7 for a couple of seconds, then try to reproduce the symbols in each group. Most people will find the left group easier to reproduce because they see it as three chunks. The right group is intrinsically no more difficult to remember, because the symbols are the Thai numerals for three, two and five. It is harder for non-Thais to reproduce, though, because it has to be seen as a large number of chunks (little squiggles, something like a rounded m, another squiggle at the top of a vertical line, a horizontal line . . .).

Size and therefore number of chunks perceived in a situation is one of the big differences between the knowledgeable person (expert, teacher, adult)

Figure 9.7 Arabic and the corresponding Thai numerals, to illustrate chunking

and the ignorant (beginner, student, child). Almost paradoxically, an expert inhabits a simpler world than a beginner because the expert breaks it into a smaller number of meaningful units. One of the clearest demonstrations of this is research by de Groot (1965) and Chase and Simon (1973) on memory for chess positions. Chess masters were far superior to beginners at recalling positions of pieces when they were arranged in places likely to have occurred in a game, but this superiority was diminished sharply when the pieces were placed randomly. The simplest interpretation is that the masters saw the natural positions as a small number of chunks with which they had become familiar. A common arrangement, for instance, is that shown in figure 9.8 where the castled king is sheltered behind its row of pawns and protected further by the knight on the Bishop 3 square. Experienced chess players picture this arrangement as a unit, knowing that as long as it remains undisturbed the king is not likely to be in danger.

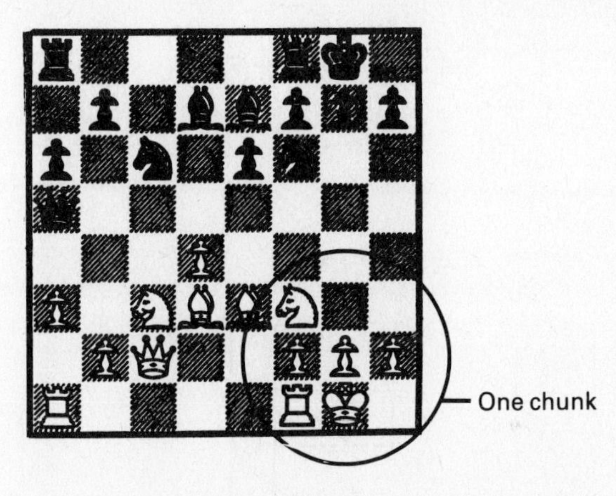

Figure 9.8 Chunking of a section of a chess position

Chunking is important in communication, and therefore in learning. When the teacher says 'Concentrated sulphuric acid is a powerful dehydrating agent', she may think of the communication, and have constructed it, as a small number of chunks. Thus: (Concentrated sulphuric acid) (is a) (powerful) (dehydrating agent). The students, who have not learned as much as the teacher, may hear it as rather more chunks: (Concentrated) (sulphuric) (acid) (is a) (powerful) (de) (hydrating) (agent). They could then be overloaded in short-term memory and fail to register the full message. If the teacher does not appreciate that the students are having to cope with more chunks than are apparent to her, she may move on too quickly for them to have time to rehearse the separate bits and form

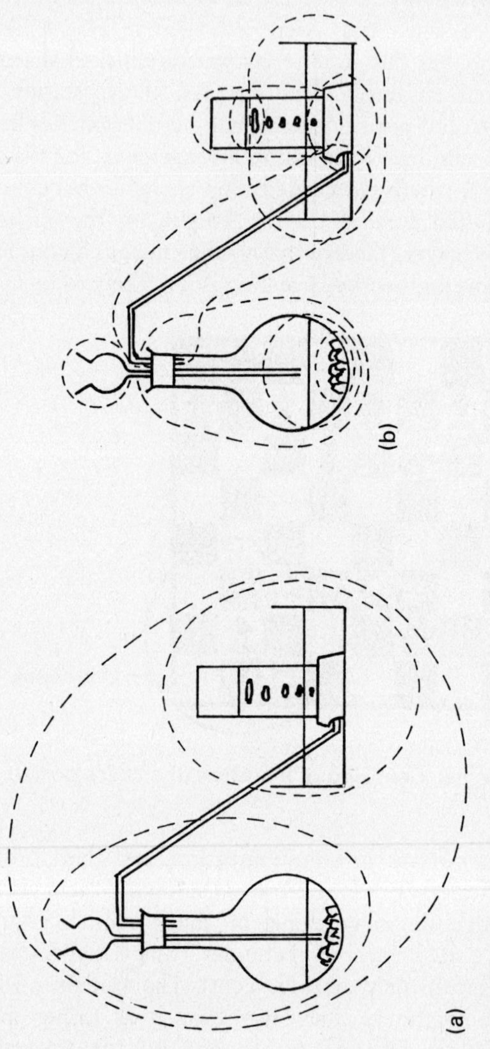

Figure 9.9 Hypothetical chunking of gas generation apparatus by (*a*) a teacher and (*b*) a student

them into new, larger chunks. If she does appreciate it, she is more likely to see it as her role to help the students to form larger and fewer chunks.

Another example is where the teacher shows students apparatus, such as that which was commonly used in schools for gas generation. The teacher may see the equipment as a small number of units, shown by the marked off areas in figure 9.9(*a*), while the students are having to cope with a dozen or more, as in figure 9.9(*b*). The teacher has not lost sight of the details of each unit; nothing has been subtracted. The teacher has acquired superordinate chunks. The principle is that the more you learn, the more integrated yet more differentiated the world becomes.

There is a small price to pay for acquiring bigger chunks. One tends to see what one expects to see, and so some fine details may be missed by the expert but seen by the beginner. In the gas generation drawings, for instance, the thistle funnel in the second drawing (*b*) does not reach the surface of the liquid, so the apparatus is incorrectly arranged. Most experts would miss this, unless they had been alerted to look for an error, as they would have chunked the arrangement without checking the details. Beginners would be more likely to record this incorrect arrangement accurately. Another example, a classic in psychology, is figure 9.10. People who are used to English sentence structure, and especially those who are familiar with the phrase 'Paris in the spring', often miss one of the 'thes' in the triangle.

Figure 9.10 Example to show how chunking through familiarity can cause misreading of detail

The occasional misreading of a detail is more than compensated for by the power that seeing things in larger and fewer units provides. By and large teachers do not train students to see things in chunks; there is potential there for improving learning.

Ability to chunk information is a learned strategy, and the act of chunking will vary with familiarity with the topic. The more you know about the topic the easier it is for you to chunk it. However, the number of chunks a person can hold may be a more fixed characteristic, and will vary from person to person. The variation is important in determining how difficult a problem will be for an individual. If the number of things you have to keep in mind at once to solve the problem exceeds your short-term capacity, then you will find the problem difficult. Johnstone and El-Banna (1986) demonstrated this by analysing chemistry problems into the numbers of steps that an unsophisticated student would take in solving them, and by measuring the short-term capacities of 471 upper secondary school and university students. After the students had attempted the problems the fractions answering each one correctly were plotted against the number of steps required, with separate curves for people with capacities of five, six and seven items. The fractions correct plunged when the short-term capacity was exceeded (see figure 9.11).

The results obtained by Johnstone and El-Banna show how short-term capacity influences problem-solving success, but this does not mean that someone with a small capacity is incapable of learning or solving problems. Such a person needs, even more than most, to develop strategies of chunking. Johnstone and El-Banna point out that the curves they found in their results imply that few of their subjects were chunking in the chemistry problems any better than an unsophisticated beginner would:

> One might even define a person, successful in *any field*, as one who has developed a set of . . . strategies which enable him to out-perform his own limitations.
>
> A beginner musician who is struggling through five-finger exercises will only be a musician of any competence when he has developed strategies for reading whole chords or even whole bars in one glance. As long as each note appears as a separate entity he will never play with competence.
>
> The *bulk* of the pupils and students in our samples . . . were working *up to* and *not above* the limit set by their own [short-term capacity]. In other words, they were not using . . . strategies to any extent. It is little wonder that most of them never become chemists. (p. 83)

The duration of short-term memory can also affect learning. It is difficult to get a pure measure of how long an item could remain there, since it could be supported by rehearsal or destroyed by interference from new items. The time may also depend on many factors, such as the nature of the item – a shape may have a different duration from a word. However, an estimate of the order of seconds, certainly less than a minute, may be reasonable. For practical purposes the time does not matter much, since we live such stimulated lives that interference from new information is far more often

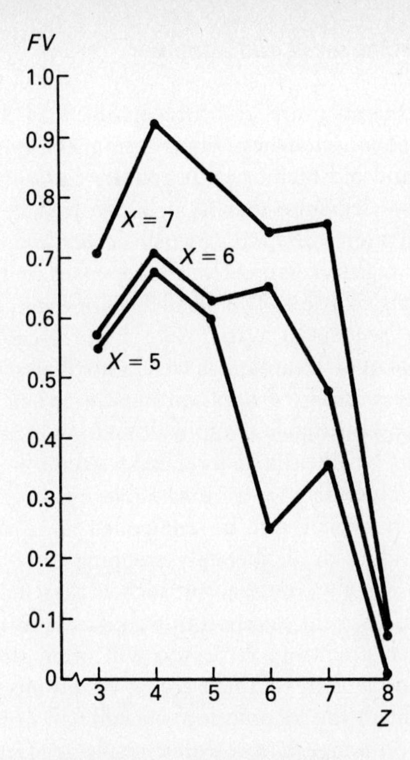

X Short-term memory capacity
Z Number of thought steps necessary to solve problem
FV Proportion of people solving the problem

Figure 9.11 Results obtained by Johstone and El-Banna (1986) demonstrating fall-off in proportions of people solving chemistry problems when the number of thought steps required exceeds a person's short-term memory capacity

the cause of forgetting an item than the fading of a trace. Even without external stimulation new thoughts will flood in, provoked by the original stimulus which thus helps its own destruction. In the face of interference, the only way to maintain an item in short-term memory is to rehearse it. Rehearsal is mental repetition, often aided by tongue movements in silent or near-silent muttering. In effect it constitutes a new input; if at any repetition an unnoticed error is made, the new form takes over from the old.

Consciousness and attention

Selection and short-term memory are crucially involved in the largely unexplained phenomenon of consciousness. Short-term memory is entered by both new experiences and old memories. It appears to be the site for mental operations on these elements; that is, it is involved in thinking. There is a parallel here with micro-computers, which accept input from the keyboard or from memory registers and perform operations on the inputs in a small number of working registers. The parallel is not exact, of course. The brain is not a simple sequential system like a micro-computer. It exhibits unconscious control over voluntary as well as involuntary acts, for instance in the flexing and relaxing of opposed muscles when someone decides to bend an elbow, and is capable of rapid alternation of attention.

Unconscious control can be illustrated by calling attention first to a common experience and secondly to a remarkable research finding. Walking is a voluntary act which can be controlled in a deliberate, continuous way, as in avoiding or deliberately stepping on cracks in a pavement or when soldiers adopt an unusual gait such as slow marching or goose-stepping. More often, once the decision to walk has been taken, the act continues without much attention. A person will often stroll while thinking of other things, and as long as there are no hazardous obstacles about will give little thought to the locomotion; virtually all of the short-term capacity is free for thinking. The research result referred to was obtained by MacKay (1973). An ambiguous sentence, such as 'They threw stones toward the bank yesterday' was spoken in one ear of a person, while at the same time the other ear received a context word such as 'river' or 'money'. Hearers tend to interpret the sentence in the context of the given word, which, however, they could not recall hearing.

Rapid alternation of attention may be more important than unconscious control in the learning of science. It seems to be the mechanism by which we manage to keep track of more than the half-dozen or so things that short-term memory can accommodate. As I write, I am conscious of my purpose, that there is a cup of tea on the desk as well as books and scattered papers, that outside the window there is a road along which cars, trucks and pedestrians are passing, that in other rooms nearby people are talking, that somewhere unfortunately close someone is hammering, and so on. It is a steady stream of information. My awareness is, however, not truly continuous for all these events. It is maintained only by scanning of the environment. MacKay's result (1973) suggests that there is an unconscious element in the scanning, which may be continuous. The evolutionary advantage of that is that we remain sensitive to events which may be more crucial to survival than the ones we are concentrating on. The conscious element is more intermittent. If I concentrate for long enough on what I am writing, my conscious scanning wili be restricted to the page and its immediate surrounds and to my thoughts. The other features I was aware of before, the traffic and the hammering and so forth, will fade from my consciousness. They will only come back if I switch my attention to them, perhaps because they reach an intensity that contrasts sharply with the level I had become used to. I can maintain them by flicking my attention back and forth between them and the task I am engaged in, though this may slow the task or dislocate the train of my thinking – I could 'lose the thread'.

Consciousness involves an awareness of identity, which has been the subject of speculation for centuries, perhaps as long as mankind has existed. Descartes' 'cogito ergo sum' (I think therefore I am), though nicely ambiguous, can be interpreted as a statement of the equivalence of thought and self. My model leaves untouched this question of self. Although it describes many of the operations of the mind in learning and doing, in the end it leaves consciousness unexplained. It is not the only model to do that. But this does not make models useless. Although the ultimate mysteries of the functioning of the mind may remain hidden, the more of its operations that can be described, the more useful a model is as a guide to practice.

Deep processing

The information in short-term memory can be processed further. Unfortunately the mechanisms of this processing are even less amenable to inspection than those of translation and selection, and have received less attention from psychologists. Recent studies by educationists have, however, enabled some principles to be formulated, which examples make clear.

Imagine a child hearing or reading that 'Black surfaces radiate heat better than white or silver ones'. This event would normally take place in some meaningful context, where what had gone before and what was anticipated to follow would affect the processing that the child does, but for the sake of this illustration let us ignore the context. Suppose now that the message has been received into short-term memory. Processing could stop at that point, because the child does not choose to go further with it, or is incapable, or is not given sufficient time before the next piece of information arrives. Choice, capability and time are central issues in teaching and learning. Much of teachers' professional conversation concerns them.

Choice

Like all organisms, humans are in dynamic relation with their environment. They are continually changing and doing things. Some of their acts are unconscious, such as movements of the muscles of the digestive system or dilation of the pupil in different lights, others are conscious but uncontrolled physical movements, like shivering or reflexive responses to pain or sudden noise, while dreams are uncontrolled mental experiences. For the most part, learning is not like those examples; it is both voluntary and conscious, and is undertaken only when it fits in with the students' goals.

Where do goals come from and how are they formed? Humans are such complex organisms that any simple answer is bound to be glib, but in tackling these questions we should obtain insights about the learning of science and the communication of meaning. We could start through behaviourism, with its central tenet that the acts that are undertaken are those that are reinforced, or rewarded. Application of this principle is complicated by the problem of what is a reward? There may not be much difficulty over that for hungry pigeons or rats, but for people it is not so easy to tell. People differ so much and have such diverse needs that what is a reward for one leaves another indifferent.

Naturally there have been attempts to codify human needs in order to make their diversity comprehensible. Murray (1938) separated 'viscerogenic' needs, those of biological origin such as needs to satisfy hunger and thirst and to maintain comfortable body temperature, from 'psychogenic' needs, which come from interaction of the individual with society and cover things like need for achievement, acquisition or affiliation. Behaviourist research with animals has related rewards to satisfying viscerogenic needs, but people hardly learn science (or much else) in a direct connection with physical drives. Learning is a social act, and is undertaken to meet social needs. There can of course be a physical connection, along the lines of 'I am learning this to get a job so that I can earn money to live well', but this is distant from the immediate press of hunger.

Students' goals that will lead them to learn science arise almost entirely from social needs. Many of Murray's psychogenic needs may apply to any taught subject as much as to science, but there are two factors that can affect science specifically, one temporary and one permanent.

The temporary factor is the status of the subject. This can make people want to study science in greater or lesser numbers, and in different ways. In the nineteenth century the curriculum was dominated by the classics, and science was of low status in schools. In *Tom Brown's Schooldays*, the boy Martin, an enthusiastic naturalist and experimenter, is a figure of fun who experiences the derision and loutish interference of his peers. A real life example is Charles Darwin, who as a child was reproved by the headmaster of Shrewsbury Grammar School for wasting his time at weekends in carrying out chemistry experiments. Despite discouragement, however, many people did study science, and, as Layton's *Science for the People* (1973) shows, did so with a commitment that produced a quality of learning now rare among those who study science because of its high status.

The fictional Martin, Darwin and the other nineteenth century students of science were motivated by the permanent factor I alluded to above. Science is a system for bringing order to our understanding of the natural world, a system which enables us to see the universe as rational and consequential without losing any of its grandeur and mystery. Whatever its current status, at all ages this characteristic allows science to meet human needs for order, control and construction. We want to understand the universe.

Such a grand goal is hard for a starving person to maintain. Maslow (1970) postulated a hierarchy of needs, with physical needs having to be met before those of belonging, esteem and self-actualization. As well as being ranked in a hierarchy, needs are related to time. While long term in the sense of being recurrent and persistent, physical needs require to be met quickly. They are more related to short-term goals than are the highest, mental desires to know, to understand and to appreciate.

People also differ in how clearly formulated their goals are and how conscious they are of them, and where the balance is between their long- and short-term goals in determining their actions. Some people appear to live entirely for the present, and are known as impulsive, or thoughtless; others have long-term plans that cause them present disadvantage. As in many things, those with a central position, a balance of short- and long-term goals, tend to have the best of it.

The issue of long- and short-term goals is relevant to learning of science. Curriculum designers and teachers may want to transmit to students the long-term goal of understanding science to a high degree, so that they obtain power over their environment and have a richer life through enhanced appreciation of the world and of the construction of reality that centuries of endeavour by scientists have created. That is a noble aim, but unlikely to be articulated by school students. They may approach it through goals such as wanting to learn more about science, to understand how things work and how plants and animals live and grow. Achievement of those aims requires the advanced learning strategies of reflection and interlinking of knowledge. But, at the same time the students may be in a context that encourages formation of short-term goals that require strategies inimical to the ones needed for the long-term goals. The students want to go on learning science, for that fits their long-term goals, but to do that they may have to pass examinations or tests that often involve recall of propositions and intellectual skills. There would be no harm in that, except that the tests are often stereotyped so that an element of drill enters the learning. Scientific laws and potentially meaningful facts are learned as strings or as propositions unrelated to experience, while intellectual skills are practised so that they can be applied automatically to a limited number of standard exercises. There is no call to contemplate the relation of things learned to episodes or to propositions that have been induced from experience, because that is time-consuming and not necessary for attaining the short-term goal of passing the examination, however helpful it may be for the long-term goal of a deep understanding and appreciation of science.

The lack of cohesion between practices necessary for short- and long-term goals in the learning of science maintains the unfortunate separation between school-acquired knowledge and the beliefs built up from experience, which has been uncovered by the studies described in chapter 5. To meet the short-term goal of passing examinations a veneer of scientists' knowledge is placed over beliefs that are often inconsistent with those accepted from the teacher. In the context of the classroom, the 'scientific' knowledge is retrieved and demonstrated in order to meet the short-term goal, but once that goal has been achieved the veneer can be allowed to peel off. Thus part of the solution to the problem of alternative conceptions of scientific principles and phenomena is to bring the short-term goals into accord with the long.

Another way of looking at goals concerns whether they are imposed or self-developed. For self-developed goals the rewards are, to a large but not entire degree, provided by a sense of psychological well-being, of self-respect and pride in achievement. These rewards are not entirely internal, because there is often an element of comparison to other people in them, knowledge that one has superior attainment. The achievement may be desired because it will bring the regard of others as well as self-approval. Nevertheless, some goals are more self-imposed than others, wherever the reward comes from. They tend to be long-term and are the mainsprings of action in a person's life. The externally imposed goals, with rewards such as a pass in a test, signs of approval, withholding of punishment, tend to be short-term, though the relation between duration and source of a goal is not exact.

With rare exceptions (saintly hermits?), there are elements of external reward in everyone's life, and we all have a mix of long- and short-term goals. The critical matter is the balance on both of these dimensions. Currently, schools are organized around external rewards and tend to concentrate on short-term goals. While they should not abandon these, better learning of science should result from a shift in the balance towards fostering long-term goals and reliance on internal rewards. Essentially that is the basis of Baird's concern (chapter 6) to train students to take more responsibility for their own learning. As long as learning is undertaken to meet externally imposed goals alone, learners will not become self-sufficient, will not acquire a permanent commitment to learning, and will not develop the cognitive strategies that Baird tried to induce of evaluating their own performance and reflecting on their learning and elaborating it.

In addition to goals, another factor in choosing to do something is the judgement of likelihood of success and the weighing of that judgement against the rewards and penalties involved. In crossing a road, the reward is usually low while the penalty, if hit by a truck, is very high, but the risk is extremely low (not in some cities) so we do choose to cross. To take a learning example, if we take a course in science with an examination at the end we have to decide whether to sit the examination. The reward may be judged high, the penalty low, the risk moderate, so we do it. If we are deciding whether to enter on a long course of study, then the reward is high but so is the penalty – the commitment of time, interference with other activities, the blow to self-esteem if we fail – so we might be put off if success is judged to be less than certain.

Judgements of the likelihood of success and of the relative weights of rewards and penalties draw on experience. Episodes are crucial in deciding whether to act or not. There is no problem in familiar situations, as there is a stock of relevant episodes in memory. When we come to a road, an examination, a new course, we can base our judgement on experience with earlier roads, tests and courses. Faced with an unfamiliar situation, we seek parallels between it and familiar ones. When a student exchanges the classroom of secondary school for the lecture hall of university, new judgements have to be made about how to behave, including how to go about learning. Whether the student changes behaviour or not depends on how similar he or she perceives the contexts of school and university to be. In familiar situations and new, it is the student's perception of the context, the rewards and penalties that are present, that influences judgement which interacts with needs to determine the performance that follows. These interactions are represented in figure 2.1 by arrows from perception of context and needs to performance.

The quality of learning can be improved through making judgement more conscious and more realistic, and through making students' perceptions of context more sensitive and supportive. In two investigations of learning styles, Baird and I (Baird and White, 1982b, 1984) found that both tertiary and secondary level students believed they were coping better with their studies in science than more objective evidence justified. While this was a good thing in that it kept them motivated and cheerful about their courses, it inhibited them from changing their learning habits and merely delayed and intensified, in many cases, the shock of eventual failure.

Judgement is a recurring process, involving not only the initial decision to act, to cross the road, enter the course of study, and so forth, but also monitoring of progress to make frequent decisions about whether to keep going and whether to change the way things are done. It is an odd characteristic of current schools that self-monitoring of learning is intermittent or even absent. One of the principles of metacognition training, which is aimed at improving learning, is to make students' judgements of their progress frequent and deliberate.

Perception of context matters in choice because it influences judgement about the rewards of learning; that is, the point of the learning. A common complaint is that students see no point in the science they learn, or that what they learn in the science classroom is not transferred to out-of-school life. Because much of the science curriculum must then appear arbitrary and useless, there is little point in learning it well. Again training in metacognition may help, through requiring students to reflect on the meaning of the propositions and skills they are taught, and to consider how they relate to other knowledge and to non-school life. As long as students restrict their view of learning to the acquisition of knowledge to pass a test, they will be content with limited styles of learning. To change to a more valuable style they would have to adopt a broader perception of the context of their learning.

On hearing that 'black surfaces radiate heat better than white or silver ones', will the student reflect on this, conjure up images of black and silver objects, recall episodes of seeing car radiators being painted black, consider why domestic electric radiators have a shiny surface, think about another statement that black things absorb heat better? Or will the student merely let the proposition slide out of memory, or try to commit it to memory by rote repetition? Largely it is a choice, determined by goals, needs, rewards, penalties, judgements and perceptions of context. But even if the student wants to process the proposition deeply and thoroughly, there remain two other factors that determine the quality of the learning: capability and time.

Capability

Processing of information can stop at short-term memory through inability, which may be permanent as a result of brain damage, or fixable through training in cognitive strategies.

Up till now, science teachers have rarely had in their classes students with marked, permanent disabilities in learning. It is likely that their experience will be broadened as shifts in social values promote integration of handicapped students into 'normal' schools. They may, for instance, meet hyperlexic students, who can read aloud fluently but are unable to carry out semantic processing. Hyperlexia appears to be an organic defect, not amenable to training, but since children with permanent handicaps have as much right and responsibility as any other to learn science, learning theorists, researchers and teachers need to find out how to help them.

Much more common than the permanently disabled are those who are innately capable but not skilled in processing. In the terms of chapter 6, they lack good cognitive strategies. Since their deficiencies are not congenital and have no physical cause, the proportion of them in a population will vary with styles of upbringing and schooling. As Baird (1986; Baird and Mitchell, 1986) has shown, this proportion can be reduced by appropriate training.

Time

Even people who can and who want to process information may be prevented from doing so if it comes in too rapid a flow. Contemplation of the meaning of a communication takes time, and that time may be longer than the pause between sentences. In reading, the rate of flow is under the control of the recipient of the information, but in oral communication it usually is not. In conversation one can ask the other person to wait a moment, but in group situations such as lectures and school classes that is more difficult. It must be a common experience in listening to a lecture to have to choose between thinking about the implications of one sentence and losing touch with the next one or two, or keeping up with what is being said without really processing it. Fortunately, speech is often more redundant than text. Repetition is employed, and appreciated, which in text would be annoying. In primary and secondary schools teachers adjust the rate of presentation to suit their judgement of the students' grasp of the information, and use questions to promote processing. Lundgren (1977) has shown that this judgement is based on the reactions of 'target' children, who are nearly, but not actually, the slowest in the class. University teachers more commonly rely on the students to put in effort outside the lecture theatre to understand the topic, and allow little opportunity for processing during the presentation. Also, as things stand at the present, they can assume there is greater motivation to process, and greater skills at it, than can teachers of younger levels.

Processing of propositions

The unprocessed sentence 'Black things radiate heat better than white or silver ones' could be stored as a string that is established in long-term memory through repetition. More likely, however, is that if it is not processed as a proposition it will be lost. Since people vary in their ability and style of processing, we can only speculate about how one individual might go about it.

Meaning of Constituent Parts

A child could begin to process the sentence by separating it into blocks, which are imaged or related to known objects or phenomena. The child might picture black and silver shapes, then some actual objects of those colours. 'Radiate heat better' may be processed by recall of an episode, general or specific, of receiving heat from a radiator, and an image might be formed, say of an electric radiator with wavy arrows coming out from it. Each block is thus given a meaning through translation of the word to an image of something known or to an episode. Imaging may not always be essential: the terms could receive meaning from propositional connections instead. 'Chancellor to bring in tough budget' as a headline in a British newspaper takes on meaning if one has propositions such as 'The Chancellor of the Exchequer is responsible for government finance' and 'The budget is the government's annual statement of what it intends to spend and where it plans to get the money from'.

Illustrative linking

Further processing can involve more and more linking of an illustrative kind. The child may remember an episode, such as being told that the black thing at the front of a car engine is the radiator. Each element that is added changes the child's understanding of the original sentence and of its constituent concepts. There is a reverberation as each new element is added and changes the child's understanding of its predecessors.

Episodes, propositions and images that are added may be thought of by the child, or be suggested by the teacher. So long as they are firmly established in memory their source does not matter, but establishment seems more certain when the child is the originator. This must not be interpreted as an injunction against teachers suggesting extra elements, but as a reason for training students to elaborate their knowledge. Reflection and elaboration are important cognitive strategies that should be fostered.

These strategies may also be employed in inventing knowledge: 'That's why light-coloured houses are better – they don't lose so much heat.'

Explanations

A special form of processing involves explanations of the statement. The child might ask why it is so, and be told, or may devise reasons for herself. Most explanations of scientific phenomena involve either logical deduction from axiomatic or directly experienced propositions, or analogies. Symington and I (Symington and White, 1983) found that children explained the fact that trees have bark most often through the analogy with skin, and the logical conclusion that it protects the inside. The explanation of why black radiates heat better is not so simple, and it may be beyond most children to devise reasons for it. That does not mean they do not understand the statement, only that their understanding of the phenomenon is not as complete as it might be. Though important, explanation is not an essential part of processing, nor is it a prerequisite for another act of processing, that of evaluation.

Evaluation

Evaluation determines the child's acceptance and degree of commitment to the proposition. The child may hear the statement and reject it, or accept it as partly true or applicable in some contexts, or accept it completely. The manner of evaluation can provide a new depth of processing. At a shallow level, the child can accept the statement because of reliance on authority: parent, teacher or textbook said it, and they must be right. Although this is a shallow level of processing, it should not be confused with the tenacity with which the proposition is held. Anyone who has tried to argue with someone who has 'seen it on the telly, so it must be true' will be aware of that. Reliance on authority is often coupled with acceding to truth in some contexts only. In the classroom the child may go along with the teacher's presentation of the world, but will activate the beliefs received there only in that context. If the teacher wants to say spiders are animals, then the child will cooperate to the extent of answering tests that way, while believing outside the classroom that spiders are not animals.

At a deeper level, evaluation involves checking the new information against old knowledge. If it conforms with it, and especially if it explains something that has been a puzzle, then the commitment to the new knowledge should be high. Thus the child may think, 'That explains why polar bears are white – if they were black they would lose too much heat', and so feel more confident about the phenomenon. On the other hand, the child may question, 'If that is so, why are the hot water radiators in this school painted yellow?'. If an objection like that cannot be resolved, faith in the proposition may be withheld.

As well as reliance on authority and checking against old knowledge, evaluation can sometimes be through direct experience. An experiment could be set up where radiations are compared from a black surface and a white surface at the same temperature. Often in schools this consists of filling two cans, one black and one silver, with hot water and measuring the rates at which their temperatures fall. I suspect this is more convincing to the teacher than to the students because the connection between 'radiate heat better' and the readings of the thermometers may not be appreciated. A better demonstration would be one in which the students feel directly the greater radiation from the black surface.

Unfortunately, many school demonstrations are indirect. I learned about Boyle's Law through apparatus that consisted of a closed glass tube joined to an open one by a U of rubber full of mercury. As the open end was raised the air in the closed end was compressed, and we were supposed to realize that this was because the pressure on it was greater now. At the time I did not follow that at all. A much better, because more direct, experience that is used now is to load a plunger with weights, or to push on it with one's hands to feel the relation between pressure and volume.

Evaluation can lead to rejection of knowledge. Rejection can occur for two reasons, either because the new statement runs counter to present beliefs, or because there are emotional reasons, such as feeling threatened by it, for not wanting to accept it. Both elements were involved in the nineteenth-century reaction to Darwin's *Origin of the Species*. The statement that man evolved from other animals could not be reconciled with biblical knowledge, and it was repugnant to many people to contemplate their descent from monkeys. Others, however, accepted it immediately because they saw it explained so much other knowledge. 'Of course!' said Huxley 'How very stupid not to have thought of that.' Both sets of reaction persist to this day. Bell (1984) found that some children did not want to accept scientists' classification of people as animals because of opinions about the behaviour of animals. 'People aren't animals; you can't call people animals.' As well as emotional rejection, there were logical reasons given, stemming from a different definition of animal: 'People don't go round on four legs'.

Resolution of conflicting beliefs

Recent research indicates that evaluation is a crucial part of processing. It is also an element that has not been recognized by teachers. Current practice places stress on relating new information to old, and on explaining statements as well as phenomena, but little or no attention is given to students' evaluation of the information and whether they are eventually committed to believing it. Studies of understanding of mechanics, using the techniques described in chapter 5, found that beneath a veneer of knowledge of Newtonian principles often lay Aristotelian beliefs (Helm, 1980; Osborne, 1980). One explanation for the persistence of the Aristotelian views is that they are more in accord than are the Newtonian principles with interpretations the students made of their earlier experience. Students have observed from early life that you do need to keep pushing objects to keep them moving, and that for the most part they do not 'continue in a state of rest or of uniform motion' unless some external force is applied. State of rest, certainly, but uniform motion? Never! All experience is against it. If you stop pushing a box across the floor, it stops. If you push harder, it goes faster. Where motion does continue after application of a force it doesn't continue long, and can be accounted for by an impetus theory. When Osborne (1980) asked why a golf ball eventually fell to the ground and stopped, a student said it was because the force of the hit was still on it for a while, but gradually wore off. Brumby (1979) found much the same reaction to evolution of skin colour. Medical students who had learned about natural selection were still prone to believe that skin colour would change in one or two generations if a family moved to a different climate. Presumably this belief is associated with personal experience of sun tanning, and the loss of a tan in winter.

Overcoming beliefs that are rooted in experience turns out to be very difficult, and teachers can never be certain that they have succeeded completely. Posner et al. (1982) set out the conditions that are necessary before an old belief is exchanged for a new: the learner must be dissatisfied with the first belief and find the new one intelligible, plausible and fruitful. It is not always simple to establish these conditions, especially dissatisfaction with the present belief, which, in order to have been maintained at all, must have served reasonably well up till now.

An obvious procedure is to demonstrate that the old belief is inconsistent with phenomena that the new one fits. For example, many elementary school children, and perhaps some older ones, have an additive notion of temperature, believing that mixing two lots of water at 10°C will give one lot at 20°C. With demonstrations and individual and class teaching, Stavy and Berkowitz (1980) apparently convinced fourth-grade students that this belief is wrong. Hewson (1982) combined direct experiences with discussions and questions that linked abstract concepts with common episodes to improve ninth-grade students' beliefs about mass, volume and density. Unfortunately, success does not always follow, partly because selective or distorted observation lets learners see only what they believe should have happened, and partly because people can believe that principles are specific to contexts so that what is true for scientific apparatus does not necessarily apply to 'real' things. An intensive effort by Gunstone, Champagne and Klopfer (1981) shows how elusive real change can be. They worked for eight days, over a two-month period, with 12 able seventh- and eighth-grade students who demonstrated initially that they had pre-Galilean beliefs about force and motion. Individually or in small groups the students weighed and dropped objects in air, water and a vacuum, measured the motion of gliders on an air track and operated computer simulations of Aristotelian and Newtonian worlds. These experiences, together with frequent discussions of the phenomena, were, it seemed, displacing the students' initial beliefs with ones more in accord with those of scientists. However, in the final session deep probing revealed that Aristotelian notions, such as velocity being proportional to force, persisted along with the newly acquired Newtonian knowledge.

The implication of this research on changing of beliefs is that learning in the form of addition of new propositions to a person's knowledge and the linking of those propositions with other elements, is qualitatively different from learning that involves reflection on contradictions between new information and existing beliefs. In the main, schools, teachers and learners cope with the former quite well, but the latter is difficult and may require greater understanding of the operation and development of cognitive strategies.

Up till now we have been focusing on the learning of propositions, such as 'black surfaces radiate heat better than white or silver ones', seeing how this involves imaging of terms, linking to other propositions and episodes, accepting or constructing explanations, evaluating and comparing with former beliefs. Essentially similar actions occur in processing of the other types of memory element, which can be treated more briefly.

Processing of strings

Although strings may be processed in the same way as propositions, people more often leave them unlinked and unevaluated. When processing of strings does occur, it often comes much later than the initial acquisition, even after many years, while linking may happen at widely separated times.

Proverbs, for instance, which one learns as strings when a child, come to be meaningful as they are linked to episodes or to propositions. Strings in science, such as 'To every action there is an equal and opposite reaction' may remain in memory as strings but also take on meaning as they are linked to propositions such as 'Action was a term Newton used for force' and to generalized episodes of pushing heavy objects and feeling their resistance or force back on you. As these links with propositions and episodes are added, the learner constructs meaning for the string.

Processing of intellectual skills

Intellectual skills are common in science curricula, especially in physics and chemistry where many algorithms have to be learned. There is a sense of achievement in acquiring a new intellectual skill. Students are pleased to be able, for example, to complete sets of exercises on falling objects that involve little more than substituting in well-drilled formulae. Many students are satisfied with that accomplishment and do not process the skill further. The skill is then accepted as knowledge which gives power over a specific form of exercise, but it is not related to a wider body of knowledge.

A central, though not essential, part of the processing of a skill is acquiring an explanation for why each operation in the algorithm is performed. Finding the velocity at a point on a curved position–time graph involves drawing a tangent to the point and calculating its slope. These operations can be learned without knowing why they work, but the skill becomes more meaningful if processing includes linking it with the propositions that a tangent has the slope of the curve at its point of contact and that slope of a position–time graph is equal to the velocity of the object; and if these propositions are justified by further propositions and images.

Another example is the skill of balancing molecular equations in chemistry. A student might have learned how to perform exercises such as: Balance the equation $O_2 + H_2 \rightarrow H_2O$. This can be (perhaps often is) learned as an abstract mathematical skill, just as words can be learned as a string. However, the skill takes on meaning, just as strings do, as it is linked to propositions, images and episodes. Helpful propositions may be:

The symbols at the left of the arrow represent the molecular arrangement before the reaction, those at the right the arrangement after.
The letters represent simple atoms.
Combinations of letters represent molecules.
H_2 represents a molecule of hydrogen which contains 2 atoms.
$2H_2$ represents 2 molecules of hydrogen.
Atoms are indestructible (in chemistry).
Molecules are destructible, alterable.
Chemical reactions involve the rearrangement of atoms to form new molecules.

Then there could be images and episodes:

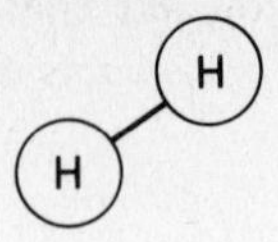

I've 'popped' hydrogen in a test tube – mist formed on the tube.

As with propositions, as each new element is linked to the skill it takes on a new shade of meaning. Processing of the skill may be intermittent, and may recur endlessly, or, as is often the case, it may cease soon after the initial acquisition of the skill.

Processing of motor skills

The points made about intellectual skills apply also to motor skills. The skill of hitting a golf ball can be acquired without any other elements being linked to it, and can remain that way. Or other elements can be added from time to time. There can be propositions – 'Keep your head down' – which can be extended by other propositions which explain why this is a good maxim. Another proposition, and image, is that a ball that is spinning sideways will curve in flight. This can be linked to knowledge of physics – Bernoulli's principle. Episodes are often linked to motor skills. One remembers the great shot one hit ten years ago, and the eagle scored at the long 15th. There may be an image of how the body should move in playing a shot. While none of these elements is essential to execution of the skill, they come to be associated with it and so are part of its processing. While they may not help execution of the skill in standard, often-experienced conditions, they may be essential in its transfer to a new situation: 'My ball is in a rut under a bush, and I have to curve it round that tree between me and the green; how will I hit it?'

Similarly, the motor skills that are acquired in science can be learned for specific tasks or for transfer. A complicated example is the reading of a scale. Children learn early to measure lengths with a ruler, both in elementary school and at home. Practice, and correction of performance, with thermometers, micrometers, ammeters, watches and protractors and other instruments, can build up a high degree of skill in measurement with almost any appliance. That transfer depends on episodes of measuring, images of scales and fractions of length intervals, and propositions about accuracy and zero errors.

A general, over-arching factor that appears to be learned from exercise of motor skills in science, whether pouring liquids, handling chemicals, preparing biological specimens, or splitting rocks, is that of precision and care. Acquisition of that general skill, which may even be elevated to the status of cognitive strategy, may rest on propositions and images as much as on extensive practice.

Processing of episodes

So many things happen around us that we cannot process them all to any great extent. Some are barely noticed. In driving to work, for instance, along a familiar route, one must be paying some attention to the events otherwise one would not arrive safely. However, it is common experience that soon after arrival the details of the experience of driving are completely lost. Although you know you must have crossed a particular road, you cannot recall whether the traffic lights were red or green when you came to it. A few events, however, are processed deeply. In a case study of my own memory (White, 1982), I found, not surprisingly, that the episodes I recalled best over a long period were those involving unusual or vivid events. What was surprising was that other properties, such as my centrality to the action, that is whether I was among the major participants in the event or just a spectator, and the importance of the event to me, and the intensity of physical sensation, had no relation to recall. My study indicated that what matters in storage and recall of an episode is how discriminable it is from other events. Driving to work is such a frequently repeated event that it is not possible to discriminate one instance from another unless something very unusual happens, such as being involved in an accident.

Processing of an episode can, by contemplating it and finding that it is similar to many similar experiences, lead to destruction of its details and its submergence in a generalized episode or script, as was defined in chapter 3. Or, its rarity or vividness permit it to be maintained as a distinct event. In either case, whether the episode is maintained as a specific experience or is generalized, processing will involve attaching it to other episodes, propositions, skills and images. The attachment requires that the episode has one or more labels given to it. For example, I have an episode which has the label 'The time we climbed the leaning tower of Pisa'. This episode also has the emotional labels of excitement and being frightened of falling. At the time of acquiring the episode I had many propositions already in memory, to which I linked it as part of its processing. These included 'Galileo is supposed to have dropped musket balls from the tower'. There were other, less obvious bits of knowledge that came into the processing. The tower is ascended by a spiral stair between the outside and an inside wall of the cylinder. The stone steps are worn, with the wear sometimes towards the centre of the tower and sometimes towards the circumference. I processed this part of the episode by seeking an explanation: the wear shifts because people tend to walk on the downhill end of the steps, and that is at different ends depending on where you are around the tower. Because of its singularity I expect I shall retain this episode for a long time, and that its recall will be stimulated by terms in the associated images and propositions, such as Galileo or worn steps.

Crucial features of the processing of episodes are the chunking of the scene and selection, which were described earlier in this chapter and also in chapter 6. Chunking and selection are part of the processing of verbal knowledge and intellectual skills, but are even more vital for episodes. Into what units do we break the scene and the action? Which of them strike us as important? The range of possibilities is greater for episodes than the other sorts of memory element, and so the meanings people construct for them can vary more. Science teaching employs demonstrations and practical work more than most subjects, so needs to be especially alert to the alternative interpretations students can make of events.

Episodes, like other memory elements, are plastic. Each time they are recalled and contemplated, details may be changed so that a new form of the episode is then re-stored. Loftus (1979) describes many instances of deliberate alteration by experimenters of people's episodes. While this is serious in legal contexts, it is not apparent that it matters much in science learning, except that prior beliefs can bring both students and teachers to believe they saw something different from what they really did. An example is provided by the probes Gunstone and I made of undergraduates' understanding of gravity (Gunstone and White, 1981). In one probe we held a light rubber ball and an iron shot put of the same diameter side-by-side, and asked the students to predict the relative times the objects would take to fall about two metres. Then they recorded what they saw happen when the objects fell. The relation between predictions and observations is given in table 9.1, and shows that there is a marked tendency to say that you saw what you believed would happen. When people are pressed about their

Table 9.1 Predictions and observations for falling spheres

	Prediction		
Observation	*Equal*	*Metal faster*	*Plastic faster*
Equal	128	28	0
Metal faster	2	10	0
Plastic faster	1	4	0

Source: Gunstone and White, 1981

report, they often become more and more convinced and vehement about its accuracy.

As well as constructing a faulty episode on the spot, by 'seeing' what one believes ought to be seen, we sometimes observe accurately but gradually amend the episode so that it becomes consistent with beliefs. This is another source of difficulty when trying to get students to abandon inappropriate conceptions. Gauld (1986) provides an informative example. Students were asked to set up a circuit consisting of a battery and globe and then to choose between four conceptions of electric current:

A The current comes from one end of the battery and is all consumed in the globe.
B The current comes out of both ends of the battery and reacts in the globe.
C The current comes from one end of the battery, some is consumed in the globe and the rest continues back to the battery.
D The current comes from one end of the battery, squeezes through the globe filament and all returns to the battery.

The students discussed their choices and then were asked to predict the relative sizes of readings on ammeters placed each side of the globe. The ammeters were put into the circuit and readings taken. The students then had to resolve any discrepancy between their predictions and observations. Three months later they were asked about their conception of current. The transcript for a student who originally chose model C illustrates revision of the episode to fit the conception: '[model C confirmed because] the meter here was more than this one but I'm not sure what they actually read . . . I think this one here was double that one . . . [not model D because] that was proven wrong by the meters' (Gauld, 1986, p. 52).

The plasticity of human memory has been important in the history of science, probably a major factor in on the one hand making science possible at all and on the other in retarding its progress. If our episodes were not amenable to social adjustment, science would be difficult. Science is an abstraction, a generalization about a variable universe. Students have to learn to overlook variations in order to accept the generalization. Their episodes from a demonstration or experiment have to be similar for common learning to occur. Science does, however, need iconoclasts. There have been many instances, for example in anatomy, where people 'saw' what authority said they should see. They were not dishonest, but came to believe that they had seen the accepted view. Hans Christian Andersen's fable about the emperor's clothes is not necessarily about dishonesty; it can describe a delusion. The small boy who saw that the emperor was naked corresponds to the creative scientist who can break away from the picture that accepted beliefs guide the rest of us to see.

Processing of images

As with the other types of memory element, processing of images involves linking. Generally, this requires that the image has a label. The image conjured up by the word 'triangle' appears because it bears 'triangle' as a label. The visual image of a triangle can be linked with another image, the auditory one of the sound of the musical instrument that carries the same label; or with an episode such as setting up the balls for a game of snooker or pool; or with a motor skill such as using a compass to construct an equilateral triangle. These associations are possible without the label, but much easier to make and to recall with it. It is a reflection on the relative importance of our various senses that we have labels for most of the things we see and hear, but few for scents. Scents, in consequence, tend to be linked with specific episodes; they rarely act as stimuli for recall of propositions or intellectual skills, except in chemistry where people learn the scents of labelled substances for which they also have detailed sets of propositions.

Science learning involves much processing of images. As well as learning many propositions and intellectual skills, we build up representations of unobservables such as electrons and magnetic fields, processes such as dissolving and burning, and generalizations such as sedimentary rock and plant cells.

The learning of images has not attracted more than a tiny fraction of the effort that has been put into investigating the learning of propositions. The relatively few research studies that have been done are mostly concerned with the effect that diagrams have on recall of verbal knowledge (e.g. Holliday, 1975), rather than the processing of the image itself. This research gap is paralleled by an odd aspect of classroom practice. For years teaching methods texts have argued against the transcription of verbatim notes, and for the construction by the students of their own notes. Science teachers appear to have heeded this injunction, but can often be seen to provide students with a diagram for copying.

Complex images require a great deal of processing to be stored accurately and permanently. For example, you would find it difficult to recall an image of the human digestive system unless you had it linked with propositions about the functions of its various parts, each of which needs a label, and to episodes of digestion and laboratory exercises such as the action of enzymes on food.

Learning of cognitive strategies

Cognitive strategies are the remaining type of memory element. They are agents of processing as well as the results of it, but they still have to be learned. Since strategies have only recently come to the attention of theorists and researchers, few of the principles that concern their learning have yet been made explicit. Clearly strategies take time to learn. Where a proposition can be acquired in seconds and an intellectual skill in minutes, an effective cognitive strategy may take years to develop.

Current efforts to study the learning of cognitive strategies use the label of metacognition or metalearning. These investigations involve direct attempts to train students in strategies such as determining the purpose of the learning, assessing the degree of understanding attained and reflecting on the relation of the topic under study to others (e.g. Baird and White, 1982b; Baird, 1986; Baird and Mitchell, 1986). There have been no investigations so far of the way in which the strategies are built up, and nothing is known yet about why some people acquire more useful sets of strategies than do others. When we know more about how strategies are acquired, a revolution in teaching could follow.

Levels of attending

The model of learning and the construction of meaning through the processing of information that has been described in this chapter allows us to identify five levels of attention displayed by learners.

Imagine a student reading a science text or listening to the teacher. The text or teacher transmits the message, 'Black things radiate heat better than silver things'. At the lowest level, the student does not select this event for attention at all, and it does not get into short-term memory. Everyone has experienced this level of attention: you have been listening, then you are distracted, and when you return to the message some time later you are aware that the speaker must have been talking for some time but you have no idea at all of what has been said. In reading one can keep looking at the book but one's thoughts take over and prevent selection of the words, which are no longer 'seen'.

At the next level, the words are selected and translated into meaningful forms and get into short-term memory, but are not processed further. They are buried by succeeding words. The student reads, in the sense of decoding the printed symbols into words, but cannot tell what they were about. This, too, is a common experience: one reaches the end of a page and is about to turn over, when one realizes that the words have made no impression; or one is aware that the teacher is speaking, without being able to recall more than the most immediate words.

The third level involves some processing, usually imaging of terms without the establishment of a large network of links to other propositions and episodes. The learner hears or sees the words, and checks out that 'black things' and 'radiate heat' and so forth make sense, but does not go as far as thinking about applications of the principle or evaluating it.

The fourth level refers to deep processing, where all of the acts of linking, explaining and evaluating are carried out. The fifth and last level is where, in addition, the learner is in full conscious control of the processing, and can extend or complete it at will. It involves the cognitive strategy of determining the purpose of the learning.

The extents to which people are capable of operating at these levels vary, but presumably are amenable to training. Poor learners are those who rarely, if ever, make the fourth level. It is not necessary, and would probably be exhausting, to operate always at the highest level, but it would be good if students' ranges encompassed it. In a way the action in many science classrooms can be seen as a struggle between the teacher who wants the students to attend at the fourth level and the students' desire, for economy of effort, to work at a lower one. The resolution of this struggle may lie in attainment of the fifth level, where the students take on rational responsibility for their attention.

Many of the actions of teachers encourage students to work at the fourth level, but as schools currently operate, few assist attainment of the fifth. The final chapter concerns the practice of teaching, and how it can be designed to promote deeper levels of attention and good understanding of science.

10

Teaching

If we could exchange a present-day teacher with one from the early years of the twentieth century, each would be bemused by differences in manners, in racial, sexual and economic constitution of their classes, and in the content of the curriculum, but they would not find their new situations so puzzling that they could not cope. The script for schools and the role of teachers within them has remained much the same for a long time.

The script predicts what we would find if we selected at random a classroom from any developed country at any year in the past one hundred. At one end of the room, an end which everyone calls 'the front' – a concept that does not apply to rooms in houses, offices or factories – is an adult called the teacher. No other adult is present. Facing the teacher in ordered rows of uniform desks or chairs are from 20 to 40 young people whose ages fall into a limited band: most are within a year of each other. It seems odd to have settled on that number of students: meta-analysis by Glass and Smith (1978) of research on the relation between class size and achievement indicates that variation in size makes little difference until the number of students drops below 16. That suggests that a more effective arrangement would be classes of varying sizes in which handfuls of students are balanced by occasional assemblies of a hundred or more in order to maintain an affordable teacher–student ratio. Universities often have that pattern, but round the world the common picture for schools is division into approximately equal-sized classes in approximately equal-sized rooms. We are so accustomed to this script for schools that it is hard to wrench ourselves away from it to see that it is not the inevitable arrangement. Even when radical (or cost-cutting) administrators in several countries built open-plan schools in the 1960s, the inhabitants of the schools tended to divide the space with screens to make enclosures of the size they were used to.

The schoolroom contains means for displaying writing and diagrams:

usually a chalkboard and, if our randomly selected room is recent enough, perhaps an overhead projector and a screen. Whatever the means, the teacher controls them and students may not use them without permission.

The teacher is the only person who talks and moves about at will. After observing many classrooms, Flanders (1970) summarized their general state: 'teachers usually tell pupils what to do, how to do it, when to start, when to stop, and how well they did whatever they did' (p. 14). He reported that students initiated very small proportions of activities in lessons. Much of the teacher's talk is in the form of questions. These are peculiar, for their purpose is not to obtain information or new views about something, as questions are used outside school, but rather to check that the students are following the teacher's line of thought or to guide their thinking along that line. The students do not ask many questions. In observing this typical example, we are aware that a lesson is going on, which consists of transmission of knowledge from the teacher to all the students simultaneously, even if some take longer than others to acquire it.

An important feature of this uniform scene is that control resides with the teacher. The students are not free to move about or to leave; they may not converse freely with each other, and may even need permission to talk with the teacher. The teacher can initiate or break off conversations at will, the students cannot. Even when students contravene the rules of the classroom they recognize their acts as misbehaviour. The script is known to them, and their presence signifies that they accept it. Real rejection is demonstrated by truancy, not misbehaviour.

As long as they remain part of the school, the students are not even free to withdraw quietly from the activities determined by the teacher, who is held responsible by the students and their parents and the school authority for managing their attention to the task in hand. The teacher is to make them learn, and selects the topic, decides the extent of knowledge that it will encompass, and directs the discussion about it. The students have little control over the course of the lesson. Their lack of control is epitomized by their writing materials. They own individual writing blocks, books, or pads, but their use is controlled by the teacher. The students are not free to write or draw whatever they will in their own books, though the teacher may inspect or write in them at any time. In several ways the participants find this a comfortable script. The teacher can plan ahead, without having to worry much about what the students want to do. The class can be treated as a unit, with conformity of knowledge as well as behaviour. The students have the consolation of reduced responsibility. They do not have to think about what to do, just obey – a simple task. If they attend to the teacher, failure to learn is either the teacher's fault or a consequence of unfortunate and unalterable lack of ability. Although it might not be the most effective script for attaining the long-term goals of the students and society, the participants' satisfaction with it makes it stable.

Another common part of the script is the textbook. For each school subject, including the sciences, the students and the teacher possess identical copies of a textbook. The very presence of identical books implies that learning is uniform. The style of the texts further implies that learning is transmission, not the construction, of knowledge.

The style of texts has changed little in the twentieth century. Lynch and Strube (1983) found that they could classify physical science texts published between 1820 and 1900 into four types, only two of which survive in any numbers today. The two that disappeared are the catechetic and the conversational styles. The example given in figure 10.1 of the former should arouse little dismay at its disappearance.

The catechetic texts presented science as a dogma, with no room for alternative interpretations. This need not have been the case with conversational texts, which could have been used to argue the merits and weaknesses of alternatives. Classic models existed in Plato's *Republic* and Galileo's *Discourses and Mathematical Demonstrations Concerning Two New Sciences*. However, the nineteenth-century texts used the conversational form for instructional monologues rather than debate. They nearly always involved a mentor and student, not two equals (figure 10.2).

Lynch and Strube note that as the catechetic and conversational forms declined, the experimentalist and formalist types took over more of the market. The experimentalist texts are similar to the laboratory manuals of the present day, and the formalist to our general texts. Both were well established, and the formalist was dominant, by the turn of the century.

In the experimentalist texts, as in present-day manuals, the apparent aim is to have students arrive at an answer to a problem. However, as we know from current practice, real discovery was rarely involved: the answer is not to be wrested from nature but from a human authority – the task is to get the answer that the teacher has in mind. As with the conversational texts, the style of the experimentalist texts implies that learning is solely transmission of knowledge, without the further act of construction of meaning.

Formalist texts dominated the teaching of science throughout the twentieth century. Although they are often more colourful now than they were in 1900, their form is essentially unchanged. As Lynch and Strube say, the purpose of such texts is to make the difficult principles of science, as set out by the great scientists, communicable to all. As syllabuses swell, the conclusions of the scientists are presented without the debate and weighing of alternative views that lay behind them. Although the better authors appreciate that science is a human construction, the format of the texts leads to the presentation of science as *the* way to see the world.

Q What is meant to be absolute motion?
A That which is measured with regard to an object at absolute rest, or where the space passed over is absolute space, – that which contains the whole universe, and which therefore cannot move.
Q Cannot we measure or estimate a motion of this kind?
A No; because we are not sure that any body is absolutely at rest. The water moves; the air moves; the earth moves; the moon and planets move; we know from the appearance and disappearance of the same spots on the disc, or face, of the sun, that the sun moves round an axis.
Q Is not that taking rather too extended a view of the works of nature?
A The wonders of creation are limited only by the power of their CREATOR, and that is far beyond what our observation, or even our imagination, can survey. System may be joined to system, and constellation after constellation of systems may revolve round centres more powerful and roll in orbits more immense. (p.51–52)
Q Whence do those variations and changes of motions proceed?
A They are usually said to proceed from differences of FORCE.
Q What do you mean by force?
A Any phenomenon that is accompanied by change in the state of body, whether that change from rest, or from one direction of degree of motion to another, is said to be produced by a force. We say, the force of wind; the force of fire; the force of gunpowder; the force of a blow; and we never see any thing in motion without thinking of the force that made it move.
Q Are force and motion the same then? (p. 53)

Figure 10.1 Example of a catechetical text (from Lynch and Strube, 1983)

The model of learning on which the presently dominant style of texts is based does not differ from that behind the texts of a hundred years ago. Learning is thought of as the direct reception of knowledge from an authority. The facts exist in the text or the teacher's head, and learning deposits a copy of the fact in the student's head. There is no appreciation that what the students already believe will influence their interpretations of the fact. The constancy with which that view of learning is held may be the key factor in the stability of schools.

One could argue that schools do not change because their function is the

Rollo seemed to be very much interested in this conversation. He had dismounted from his father's knee, and stood by his side, listening eagerly. His mother, too, was paying close attention. As for Nathan, he sat still; though it is not by any means certain that he understood it very well. 'Let us suppose', said his father, 'that the mass of lead, as big as a load of hay, is fastened to one end of a stick of timber'. 'That would not be strong enough to hold it', said Rollo. 'Well, then, to a beam of iron, as large as a stick of timber', rejoined his father. 'O', said James, 'you could not get such a big bar of iron'.

'No', replied his father, 'only an imaginary one; and that will be just as good as any. Now, suppose the great mass of lead is fastened to one end of this bar, and another one, just like it, to the other end, to balance it. Now suppose that the lower end . . .' Rollo began to laugh aloud at this idea, and looked very much interested and pleased. 'O, then I wish there was no gravitation', said Rollo; 'I do, really'.

'But, then', continued his father, 'if you should get up into the air, you could not get down again'. 'Why not.' said Nathan, beginning to look a little concerned. 'Unless', said his father, 'you had something above you, to push against, so as to push yourselves down. You would be just a boy in a boat, off from the shore, and without any paddle or pole. He could not get back again'. 'We might tie a rope to something', said James, 'before we went up, and so pull ourselves down.' (p. 148)

Figure 10.2 Example of a conversational text (from Lynch and Strube, 1983)

same now as it was in 1900, but that seems hard to defend when one considers the remarkable changes in society during the twentieth century: the population explosion, the rapid rise in prosperity, the changes in values for authority and for the role of women, the shifts in national power, the technological revolution, the concern for the environment, the possibility of self-destruction of the species. Instead, it is likely that the function of schools has changed, and that many of the stresses associated with them are a consequence of a mis-match between the new function and the old form. The rate of change of society calls for a change in the quality of learning. Transmission of well-established beliefs suits a complacent, static society, but dynamic conditions require alert, independent, lifelong learners. As much of the change in our society is related to science, we need particularly a change in the quality of learning of science. That, of course, requires a change in the view of how science is learned, which in turn has consequences for the practice of teaching.

My theme throughout has been that learning is not the simple absorption of knowledge but the construction of meaning through the individual's relating things seen and heard to things already known. Learning is active, not passive. This principle creates a dual role for the teacher, only one part of which is present in the current script for schools and even then in a somewhat different style. That part is the responsibility to put appropriate information in the way of the learner, and to arrange it in a form that maximizes the learner's chance of understanding it. The other part, ignored in most current practice, is to promote the learner's ability to construct meaning. These two parts of the teacher's role are intertwined. It would be a travesty for the teacher to try to separate them, devoting one day to imparting information and another to training the construction of meaning. I shall try to illuminate both in the sections that follow on content, selection of information within a topic, care over language, sequence and pace of presentation, questioning, use of the laboratory and teaching style.

Content: what is to be learned?

Children growing up in an American city need to know different things from those in an agricultural region of Asia, and what both need to know now is not what was needed last century nor what will be useful late in the next. Content varies with circumstance, so instead of prescribing details of topics it is better here to consider principles that apply generally.

A major difficulty in determining what should be learned is that school science courses have to serve two purposes that are not easy to resolve. Secondary education for all came late, so that the first science courses were designed to prepare an elite for further specialist study. When secondary schools broadened their intake, another need appeared, of making science courses useful to all. The two demands, of science for specialists and science for all, are with us still. Societies need specialists to keep a technological civilization operating and to advance scientific knowledge; they also need an informed populace that understands the function of science and has a balanced cultural development. Unless the mass of people understand the possibilities and purposes of science, they will not support it with votes, or in other ways of expressing opinions, for committing resources, materials or people, to it. The belief that science is valuable, if only for selfish reasons of national standing, led to the enthusiasm with which funds were devoted in the United States and Russia to the exploration of space. The mass of people have to decide questions of ecology, and questions of ethics in issues such as in vitro fertilization and cloning, and need some understanding of science to decide well. People have to appreciate the potential and the limitations of science, that it is rational and not magic. Their cultural development is involved through the pleasure and richness of life they gain from comprehension of the physical universe.

There has to be a balance between the two demands of science for all and science for specialists. This balance can be seen as a tension between the observational and the conceptual, or, in terms of the model of learning, between episodic and semantic knowledge. Because the function of secondary schools to prepare students for further specialist study still dominates, most science courses emphasize the conceptual, the semantic, the propositional side of knowledge. One way of coping with the need to make school sciences useful for all while not abandoning the specialists is to shift the balance more to the observational, or episodic side.

What does shifting the balance imply? Consider what specialists need from school science, the basic principles on which further courses will build. In physics, these are presumably the definitions and interrelations of fundamental and derived quantities, and their measurement. These notions can be given an experiential or episodic base by having students measure things, and measure them in several ways: lengths with rulers, chains, theodolites; temperatures with mercury, alcohol, gas, resistance and colour thermometers; density by weighing and measuring lengths or finding volume by displacement; forces by extensions of springs or accelerations produced. Future specialists in chemistry must comprehend how the electronic structure of atoms gives elements their properties, and how elements combine to form compounds. If these things were taught by showing the students the elements and having them handle them wherever possible, by getting them to construct compounds and to separate mixtures, their knowledge would have roots in experience and not be so conceptual and thus unsuitable for non-specialists. Even more can be done in biology than in physics and chemistry to prepare future specialists through experience with materials and practical techniques. That has been demonstrated in several courses, for instance the Biological Sciences Curriculum Study (BSCS), which contain a greater practical component than their physical science counterparts.

The other thing that all scientists need, and that all people can profit from, is the ability to apply scientific method. Part of that method is reflection on observations: framing questions to oneself about why something happened. Figure 10.3 is an example of a good scientist at work.

Students need practice in reflecting on observations, so many of the episodes they experience should require them to frame questions. Instead of

Frozen spoons and microwaves make heated brew

By SURENDRA VERMA

WHAT do microwave ovens, frozen spoons and espresso coffee have in common? The relationship between these three disparate objects will escape you unless you have a scientific bent like that of Robert Apfel and Richad Day, both of Yale University in the USA, and Anthony Parsons, of the University of York in the UK.

Apfel and Day assert that when an observation in the physical world appears to be paradoxical, running counter to one's intuition, experience and learning, it usually provides an opportunity to achieve some scientific insight. Such a case began with the observation by Day that a large spoon that had been kept in the freezer would initiate boiling in water that had been heated in a cup in a microwave oven, which Apfel suggested might occur because crystals of ice initiate boiling in overheated water.

Water is said to be overheated or superheated when its temperature at atmospheric pressure exceeds 100 degrees Celsius without changing to steam. Apfel and Day found that the temperature that could be achieved by tap water in a beaker heated on an oven-top burner was 100.75 degrees. But when tap water was heated in a microwave oven, the temperature was in excess of 110 degrees.

Apfel and Day used several kitchen objects — a stainless-steel serving spoon, glass, a screwdriver and a plastic utensil, which were all cooled in a freezer —to check their hypothesis that ice crystals initiate boiling in ordinary tap water and found they all behaved in the same way. Ice cubes — made with degassed water as well as with tap water saturated with gas — also triggered boiling if the temperature rose above about 102 degrees at a temperature of about 110 degrees, ice could initiate boiling for three to four seconds.

In a recent issue of the British journal 'Nature', the Yale University researchers explain the mechanisms of the phenomenon they observed as follows: "The microwave oven heats the water directly and the container indirectly, the opposite of what happens when heating by conduction with a burner. Therefore, the internal temperature can rise above the boiling temperature at positions away from the container's surface which tended to provide sites for boiling. Evaporation at the free surface does not occur rapidly enough to cool the bulk of the sample.

"When an ice cube is introduced, its creviced crystalline structures, which can hold gas or water vapor, stimulate film boiling on local parts of the surface of the ice cube. The vapor film acts as a partial thermal insulation, slowing down the heat transfer between the ice cube (at about -15 degrees) and the water (at 110 degrees). When the water temperature immediately around the ice cube drops below 102 degrees or so, the observed boiling ceases."

Apfel and Day's report led Parsons to recount — in a letter to 'Nature' — his experience of using his laboratory microwave oven for preparing hot drinks (a practice sure to get him into hot water with the safety officer, he was quick to add).

Parsons writes: "I had placed a cup of tap water in the microwave and set to boil but at first was prevented from making the drink by an experiment that required attention. On the third occasion of activating the oven, I responded to the timer bell by rushing to the oven, removing the cup and stirring my hot chocolate powder into the water which promptly erupted into a foaming cascade."

He interprets this dramatic event by saying that the water had become superheated and that the chocolate powder provided nucleation sites for the water to boil. As well as offering supportive evidence for Apfel and Day's study, Parsons says, "We think that a practical application of this phenomenon with coffee might be the production of energy-efficent espressos!"

Microwave ovens, frozen spoons and espresso coffee — perhaps they do have something in common. They have at least linked the thoughts of three scientists as disparate as Apfel, a mechanical engineer, Day, a paediatrician, and Parsons, a biologist.

Well, this is the kind of stuff science is made of — and good cappuccinos.

Figure 10.3 An instance of a good scientist at work (from *The Age*, 10 November 1986)

routine exercises to see whether momentum is conserved in collisions between carts ('if it weren't, we wouldn't be asked to find out'), students could be shown a phenomenon and required to come up with questions about it. Those questions could be followed up later. For example, converging and diverging lenses could be placed near a wall. On noticing that the wall is darker behind the diverging lens, a student might ask, where has the light gone? Another might ask, what sort of picture would you get behind the converging lens if it had twice the diameter? Twice the focal length?

Another part of scientific method is varying one factor to study its effect while holding all other factors constant. That procedure could be taught more deliberately and explicitly than it often is.

None of the things I have mentioned – the basic principles of science, measurement, observation, reflection, the scientific method – are irrelevant to specialists, while they would help non-specialists to understand their worlds. Unfortunately the minutiae of scientific conceptions have squeezed measurement, observation and scientific method out of most syllabuses. Often these minutiae involve intellectual skills of identifying values of quantities and substituting them in formulae. While that is an important part of science, it should not dominate school science to the extent that it does now.

As an example of minutiae, many school physics courses include Faraday's law of induction, quantified as the Faraday–Neumann Law $\varepsilon = -d\varphi/dt$. Teaching often concentrates on training students to apply that formula to exercises such as that in figure 10.4. The students are practised in algebra, not in physics. All that people need to learn in school about induction is how it can generate electricity, and that their intuitive feelings about it are correct: bigger magnets, more turns of wire, and faster rotation all lead to bigger e.m.f. Those feelings can be checked out by qualitative experiment. Concepts like flux density and the quantitative use of the formula can be left to tertiary courses.

In chemistry, Fensham (1984) suggests that instead of a conceptual presentation illustrated by experiments, school courses should concentrate on the things that chemists do: analysing, synthesizing, purifying, studying structures, searching out properties and controlling reactions. A syllabus organized on those lines would involve much doing, with students learning chemistry through exercises such as those Fensham suggests of cleaning clothes and other materials, making and repairing with fibre glass, distinguishing between common substances like kerosene and turpentine, choosing among detergents, stemming corrosion, developing films. The unifying principles that are introduced must flow out of the practice, and should be presented only in as much detail as their relation to everyday

16. A U-shaped wire (Fig. 23–30) has a movable wire *AB* connected to it. This arrangement is in a uniform field perpendicular to, and into, the page.

(a) If the magnetic field strength is 40 newtons/amp-m, what is the induced EMF (in volts) when *AB* is in the position shown and moving at 20 cm/sec? Calculate this first from the rate of flux change and then from the magnetic force on the elementary charges in the wire.
(b) What is the induced EMF when *AB* is 5.0 cm from the left end and moving at 20 cm/sec? At 10 cm/sec?
(c) At what rate is energy fed into the loop when *AB* moves at 20 cm/sec and the induced current is 2.0 amperes?
(d) If the induced current is 2.0 amperes, what force is needed to move *AB* at 20 cm/sec?

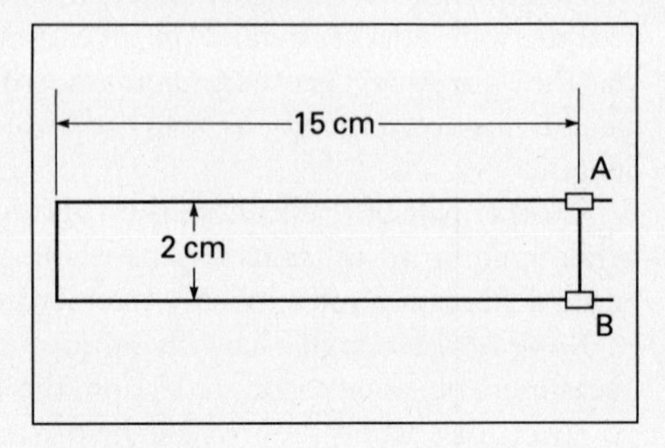

Figure 23–30
For Problem 16.

Figure 10.4 Example of an exercise on Faraday–Neumann Law (from PSSC 3rd edn, reprinted by permission of D. C. Heath and Company)

applications of chemistry warrants. The same approach could be adopted in biology and Earth sciences. The students would concentrate on observing and describing, looking for parallel structures in different species and land forms. Only then would the general principles be taught.

In terms of the model of learning, this approach will provide in advance the episodes to which verbal information can be linked so that meaningful propositions are stored. It might be objected that there is practical work in courses now that illuminates existing propositions and increases their meaningfulness retrospectively. While this is a possibility, it suggests that there is greater use of the cognitive strategy of reflection than currently appears to occur. The difference between experience as a basis for processing and experience as illumination is important. The model implies that, though some practical experience is better than none, experience gained beforehand is superior to experience gained after learning the verbal parts of a topic.

As well as affecting the quality of understanding of science, this approach will alter students' habits of learning. They will develop different cognitive strategies from those they acquire from the long-fashionable style of teaching based essentially on verbal transmission and exercise of intellectual skills. The strategies would have to be fostered, but there is more likelihood of students learning to reflect on observations, to evaluate the truth and meaning of propositions, and to determine the point of a lesson and its relation to other topics, if the verbal information builds on the episodes they have acquired.

Selection: what to say

Even if much more experience is introduced to science courses, a great deal of learning will still follow from things the teacher chooses to say. A simple notion of communication represents the teacher as drawing a fact out of his memory and speaking it so that the learners can hear it and tuck it away in their memories, but the model of learning shows it as a much more complex process. The teacher has to decide which propositions in his head are best to communicate next, has to frame them into words and utter them, and each student has to process the words and construct a meaning for them.

The teacher cannot hope to transmit much meaning if she says only: 'Today we learn about vectors and scalars. Vectors have magnitude, direction and sense. Scalars have magnitude only.' The teacher wants the students to form propositions from the last two sentences. While it would be easy to have the students acquire them as strings or as isolated propositions, that would be poor teaching, as hardly any meaning would have been shared. For the students to be able to construct deep meaning, the teacher will have to give them more than the two content sentences above, and will have to bring to their attention the relevance of things they know already to this new knowledge. Skill in selecting the additional statements, from all the strings, propositions, skills, images and episodes that the teacher knows, is at the heart of good teaching. The teacher might, for instance, have thought about vectors as an $n \times 1$ matrix or as a subset of tensors, but, although these are part of the meaning of vectors for her, she has decided not to transmit them because she judges that the students would then have so many new facts to relate to each other and to old knowledge that some would be forgotten or poorly related. Poor teaching can occur through trying to give too much knowledge as well as by giving too little.

How to achieve the correct balance can hardly be described. Examples can be given, but general principles are elusive. It is unfortunate for those new to teaching that judgement of what to include cannot be taught as principles; they have to acquire it from reflection on teaching sequences that went well and those that did not. That reflection becomes more productive as, through experience, they become familiar with the sorts of things students of a particular age might be expected to have done and to know. Until they acquire that familiarity, it is almost impossible for teachers to select well what to say from their own knowledge.

Communication: how to say it

The teacher, then, selects some knowledge, and speaks or writes a verbal form of it from which the student has to construct meaning. The model of learning emphasizes that that meaning depends on the person's existing knowledge: different people can construct different meanings from the same message, and the one person may construct a different meaning if placed in a different context. Indeed, without a context many sentences are ambiguous: 'the bat flashed through the air' (sport? flying mammal?); 'the shell was hard to lift' (artillery? giant clam?); 'acid corrodes' (LSD and society? sulphurous acid and marble?). There are many things that the teacher can do in framing sentences to aid or hinder students' constructions of meaning: sentence construction, economy of words, use of qualifying and connecting words, and establishing meanings of nouns and verbs.

Sentence construction

There is nothing special about science teaching as far as sentence construction is concerned, but because it is so important, at least a brief comment is called for here.

Speakers tend to leave sentences unfinished, so that in deriving a meaning the hearers not only have to relate the words to their knowledge, but also have to construct, from the context, their knowledge and their acquaintance with the speaker, what the rest of the sentence might have been. To assume that meaning has been shared in any communication is risky, but is additionally hazardous in incomplete sentences.

Failure to complete sentences is less of a problem when the participants are well acquainted with each other and with the topic. In exceptional cases people can be communicating well when an outsider might find it difficult to make any sense of what is being said. The Watergate tapes – recordings of conversations between Richard Nixon and his aides concerning the Watergate scandal – are an example. Transcripts of these conversations (see figure 10.5) are almost incomprehensible, even to people who are familiar with the publicly known sequence of events; but those taking part appear to have been satisfied that they understood each other.

The Watergate transcripts are of conversations, and some unfinished sentences may be the result of interruptions by speakers who wanted to get their view considered or who assumed that they knew what the other

Appendix 14. Meeting: The President, Haldeman and Ehrlichman, EOB Office, April 14, 1973. (8:55–11:31 a.m.)

This is the day, according to Ehrlichman, on which President Nixon received his first detailed report on Watergate, based on a three weeks' inquiry. It implicated Mitchell, Dean and Magruder. Edited transcripts of four meetings and three telephone conversations in the course of that day were released by the White House. This first meeting took up a wide range of topics but concentrated on how to deal with the three named in Ehrlichman's inquiry. And the pros and cons of the payments of hush money to Hunt and others were still being discussed. "Here's your situation," Ehrlichman told the President. "Look again at the big picture. You are now possessed of a body of fact. And you've got to—you can't just sit here. You've got to act on it."

(Material unrelated to Presidential actions deleted)

P—Did you reach any conclusions as to where we are.

E—No conclusions. Dick Wilson, I think, has an interesting column this morning (unintelligible). It's all a money problem (Unintelligible) Well, yes—

P—Wilson's in the Star.

E—(Unintelligible)

P—So what—?

H—(Unintelligible) is really the essence of this whole thing is too much money. Too much was spent. And so I—

P—Yeah. My point, everybody—

H—No not everybody. Let's say, one group, pieces that (unintelligible) has on that side and more like (unintelligible) says that his, you know, solving Watergate doesn't take care of it.

P—Lots of people, I think want the President to speak out on the whole general issue of money and campaign and all that.

E—Generally, but he gets specific on this. He says also (unintelligible)

P—Is that what you think, go out and make a speech?

E—I'll tell you what I think. I think that the President's personal involvement in this is important. And I know—

P—I don't think it's a speech. Well, that's a point. I think there are other ways you can get at it. Now I was thinking of the—before we get into that though let's get back. I'd like to go in, if I could, to what your conversation with Colson was and in essence, what did he and the lawyer tell you about?

E—That visit was to tell me that Hunt was going to testify on Monday afternoon.

P—How does he know that? How does he get such information?

E—Undoubtedly through Bittman.

P—Right.

E—or Bittman to Shapiro(?)

P—Now why is Hunt testifying? Did he say?

E—He said. I'll tell you what he said and then I'll tell you what I think the fact is. He said Hunt was testifying because there was no longer any point in being silent. That so many other people were testifying, that there was no—he wasn't really keeping any (unintelligible)

P—Yeah.

E—It wouldn't add much. My feeling is that Bittman got very antsy.

P—Why?

E—This grand jury started focusing on the aftermath and he might be involved.

H—Exactly.

P—What did he say?

E—He went to the U.S. Attorney and he said, "Maybe I can persuade my client to talk."

P—What do Colson et al, Colson and Shapiro, think we ought to do under these circumstances? Get busy and nail Mitchell in a hurry?

E—Yes.

P—How is that going to help?

E—Well, they feel that after he testifies that the whole thing is going to fall in in short order.

P—Right.

E—Mitchell and Magruder will involuntarily be indicted. Both will say you have lost any possibility of initiative for participation in the process.

P—What does Colson want us to do?

E—He wants you to do several things. He wants you to persuade Liddy to talk.

P—Me?

E—Yes, sir. That's his—I didn't bring my notes, but, basically

P—Oh. Last night you didn't mention that.

E—I thought I had.

P—Maybe you did, maybe you did. I would need to let—bring Liddy in and tell him to talk?

E—You can't bring him in. He's in jail.

P—Oh.

E—You would send, you send word to him, of course through a spokesman or in some way you would be activist on this score.

H—There's no, that doesn't involve any real problem. As Dean points out, he is not talking 'cause he thinks he is supposed not to talk. If he is supposed to talk, he will. All he needs is a signal, if you want to turn Liddy on.

P—Yeah. But the point—that Colson wants to call the signals. Is that right?

E—He wants you to be able to say afterward that you cracked the case.
P—Go ahead. What else?
E—Well I forget what else. You remember, Bob, when I was busy (unintelligible). He feels that the next forty-eight hours are the last chance for the White House to get out in front of this and that once Hunt goes on, that's the ball game.
P—But you've got to be out in front earlier?
E—Well,—
P—But, I mean to go public—
E—Either publicly, or with provable, identifiable steps which can be referred to later as having been the proximate cause.
P—He's not talking because he thinks the President doesn't want him to talk? Is that the point?
E—He's—according to them, Mitchell's given him a promise of a pardon.

Figure 10.5 Transcript of Watergate tapes (from *The White House Transcripts*, 1973)

person was going to say. In classroom talk, where the difference in status between teacher and pupils is even greater than that between a president and his aides, the teacher's speech is rarely interrupted directly by the pupils, though their behaviour often leads the teacher to break off a sentence to issue a command. The pupils, however, are in a weaker position. Not only do they have little chance of initiating a conversation with the teacher, they are also vulnerable to interruption by the teacher who does not appreciate how much there is to gain from listening to the students' expression of what they understand. Teachers feel so much pressure to get on, to cover content, that they are often impatient with halting construction of sentences by students. Good teachers restrain that impatience and listen.

A teacher with the habit of leaving sentences incomplete communicates poorly. As bad is the one who buries meaning in a rambling discourse of long sentences in which the grammar is complicated and often wrong. Indeed, such sentences may be even worse than incomplete ones, because the effort to follow the sense precludes the students from making any construction at all. The following excerpt from a lesson on density is comprehensible in print, but must have been difficult for the students to understand at the time:

The water will spill you see maybe if you when I lower the block the lead block come closer and don't crowd lead is a very dense material partly because each atom is heavy and they are closely packed and it displaces its own volume of water and when it does that the lead atoms push the water atoms or really I should have said molecules to one side so the water rises up and spills over the top.

In order to restrain the diseases of unfinished and confusing utterances teachers should check their performance occasionally by recording a lesson and scoring its clarity.

Economy of words

Another potential source of misunderstanding in communication is that the need for economy in speech or writing can lead all of us to say things that are not strictly what we believe. This is a particular problem in science teaching, where many of the concepts are abstractions that do not fit the real world exactly and where many of the principles are marked by exceptions. The teacher might say, 'Mammals are born alive; reptiles hatch from eggs', which is a useful distinction but misrepresents the exceptional cases of monotremes and viviparous snakes. As it would be tedious to be forever listing exceptions to principles, and would indeed inhibit formulation of almost any principles at all, we are bound to go on saying these things that we know are not absolutely correct.

Receivers of information are usually not familiar enough with the topic to know about the exceptions, and so build up incorrect deductions from the principles. 'Birds fly; fish swim', therefore penguins are fish. Similarly, people confuse bats with birds, whales with fish, turtles with amphibians. Once a deduction has been made it is, as observations by Symington and White (1983) suggest, hard to eradicate. Sometimes constant correction and publicity given to exceptions has an effect, so although children learn that birds fly they also hear the phrase 'flightless birds' and see ostriches, emus, kiwis and penguins singled out in books and in zoos. More often, though, correction is not prominent and a misleading generalization is absorbed. Examples are the faith in the universality of Ohm's and Hooke's Laws engendered by concentration on limited sets of materials and physical conditions.

Economy of words creates a difficulty for teachers. To forever emphasize exceptions may be to obscure the set of principles the students should acquire. People want simple answers. The exceptions are important, however, not only in their own right but also because they transmit a message that is often missed in school; the message that science is an abstraction, created by humans, that provides an imperfect fit to the world.

The solution, as in many things in education, lies in balance. While the principles are emphasized, the exceptions must be noted. Young children, who almost certainly come to class with the proposition 'birds fly' already formed from observation, now should learn 'Nearly all birds fly', and draw, in order to acquire a useful image, a picture with two parts, one labelled 'lots of birds that can fly' and the other 'a few that can't'. Older students could learn Hooke's Law as '*Over a short range for some materials*, extension is proportional to force', and should learn about the restrictions of range and material.

Qualifiers and connectives

Because science is an abstraction that does not fit the universe exactly there are exceptions to its principles, which consequently often need qualifying. The words 'nearly all' qualified the principle 'birds fly', and invite their hearers to construct a different meaning from the unqualified principle. 'Most chlorides are soluble in water' for example could stimulate a different construction than it would without the 'most'; an alert student would think 'Most – so there are some that are not. Why? What makes them insoluble?'. Unfortunately, many children do not understand the implications of qualifying words like 'most', 'generally', 'often', 'many', 'nearly', 'some', 'usually' and 'frequently'.

Another class of words that matters in science education is conjunctions. Some, such as 'and', represent a simple logical relation and are extremely common in speech and writing, so people grasp their meaning early in life, usually before they are old enough for school. Others such as 'consequently' and 'nevertheless' are rarer, and because they represent subtle, complex relations their meaning is acquired more slowly. They tend to be part of adult, not child, speech, and then often only of adults whose trade is in words or who read a lot. Gardner's discovery that many secondary school students cannot use correctly the less comon logical connectives (1977a) is understandable. He gave items such as those shown in figure 10.6 to more than 16,000 children in grades 7 to 10 in Australia. The percentages selecting the correct alternative are given for several connectives.

Unfamiliarity with logical connectives inhibits and distorts processing. Either the students spend so much time working out what the teacher meant by saying 'On the other hand' that the next few sentences are missed, or, more usually, the students miss the point of the connective and so construct a meaning different from the one that the teacher intended to convey. Erroneous constructions follow from misunderstood connectives. Accurate and quick interpretation of connectives is needed for a student to construct propositions that have similar meanings to those of the teacher.

What then are teachers to do when students come to their classes with a poor grasp of qualifiers and logical connectives? Should they avoid all but the commonest forms? That can hardly be recommended, since it would require remarkable self-discipline, adding to the difficulty of an already strenuous task, and could lead to an impoverishment of vocabulary and expression that would amount to mental mutilation. In any case, writers of

Hydrogen is a very light gas _____ other gases such as carbon dioxide and hydrogen.

A according to B on the basis of
C as a result of D except E compared with

	Year			
	7	8	9	10
% correct	69	75	83	92

Ionic salts dissolve in water to form solutions which conduct electricity. Sodium chloride and potassium chloride are both ionic salts;
_____, solutions of both salts will conduct electricity.

A alternatively B on the other hand
C consequently D comparatively E compared

	Year			
	7	8	9	10
% correct	20	40	43	57

Experiments with static electricity seem to work best in dry conditions. On Tuesday the electricity experiments worked quite well
_____ it was a damp morning.

A because B provided that
C as though D unless E considering that

	Year			
	7	8	9	10
% correct	67	68	80	83

The hummingbird flaps its wings hundreds of times per second.
_____, the eagle flaps its wings very slowly.

A In other words B Otherwise
C In turn D In contrast E Indeed

	Year			
	7	8	9	10
% correct	30	50	55	70

Figure 10.6 Examples of items testing understanding of qualifiers and connectives (from Gardner, 1977b)

texts will not observe the same discipline, and the students must come to be independent of their teachers, able to learn from other sources. Balance is important: eschewing connectives and qualifiers inhibits development of the students' language and prevents expression of important relations; lavish, uninhibited use, without the support of training, shifts meaning to beyond the students' grasp. Teachers have to use connectives and qualifiers, even though they are aware that they cause difficulty. Some other procedure must be followed to help students understand them.

Science teachers can ask their colleagues who teach English to give special attention to qualifiers and logical connectives, but what should they do in their own classrooms? Now and again they could explain the meaning of the words, being explicit for instance about why 'often' has to be part of a particular proposition or about the cause and effect relation expressed by 'consequently'. In those examples, students could be asked for the exceptions that make 'often' necessary, and asked to identify which is the cause and which the effect that 'consequently' spans. Occasional exercises in which students create sentences around connectives like 'because', 'therefore' and 'however', or with qualifiers like 'most' or 'often' could be useful. Another exercise requires students to point out differences in meaning that presence and absence of qualifiers create, or that are produced when one connective replaces another. For example, students could explain the difference between 'Non-metals are poor conductors of electricity' and 'Most non-metals . . .', or between 'Calcium carbonate is insoluble in water' and 'Calcium carbonate is barely soluble at all in water'. A connectives example is 'The liquid turned litmus red, because it was an acid' with 'The liquid turned litmus red, therefore it was an acid'. Other useful training is obtained by turning the test items illustrated in figure 10.6 above into exercises: '"The platypus lays eggs, (   ) it is a mammal." Fill the gap with the best word from so, because, nevertheless, consequently.'

As well as training students in the use and interpretation of qualifiers and logical connectives, teachers need to be sensitive to their importance in conveying meaning, and should check whether the students comprehend their implication and should probe whether the constructions the students have made are reasonable.

Nouns and verbs

Presumably the dawn of language involved naming, so that concrete nouns and verbs for actions such as run, kill, and eat came long before there was a need to express the subtle meanings covered by qualifiers and logical connectives. Despite the apparent simplicity of nouns and verbs, however, they too need careful use in science teaching.

Nouns (and verbs) are labels given to concepts, and in chapter 4 I described how people will have different elements of knowledge associated with a label. Also, they will differ in the cases that they classify as instances of that label. Sometimes the cases a person includes shift with the context. Science teachers at home may readily apply the label 'metal' to steel, solder, or brass, when next day in class they are careful to exclude them, naming them as alloys and restricting metal to a set of elements.

When people associate different cases with a label, or interpret the meaning in different contexts, problems occur in communication. The teacher says 'Animals cannot make their own food', a principle that students are meant to contrast with the proposition that most plants can convert carbon dioxide and water to carbohydrates. However, a young student who does not include humans in the class of animals may interpret the teacher's statement as '(non-human) animals cannot make their own food because they have no hands or stoves or kitchen utensils'. Note that this interpretation also involves a difference in meaning for the simple word 'make'.

Perhaps because there are fewer of them, verbs are less precise than nouns, so that the difficulties in communicating meaning are even greater with them. A teacher demonstrating a principle in mechanics is quite likely to say 'Now I put a force on this ball', or 'The ball has a force on it'. The verb 'put' has connotations of transfer, and 'has' of possession, of some substance. Such connotations may lead students to construct meanings for the statements that are not what the teacher intended. They may think of force as a reified quantity, a conception that explains the statement, mentioned in the previous chapter, by the student who said the golf ball eventually stopped because the force of the hit wore off. To him, force was like a fuel which was consumed in motion. Of course, that interpretation stems from analogy with a familiar system, but it is fostered by language.

People build up their meanings for both nouns and verbs from things they are told and from their own observations. In both cases idiosyncratic meanings can be constructed. Suppose a child reads 'Amphibians are animals that spend some of their life on land and some in the water'. While correct, this is not a sufficient definition of amphibians. The distinction between necessary and sufficient conditions is quite subtle, however, and even sophisticated scholars are sometimes tricked by it. The child, in reflecting on the statement and trying to make it concrete, may think of animals that fit it, such as turtles, crocodiles, beavers, otters, or platypuses, and so will develop an incorrect understanding. Another example is the way children come to restrict 'animal' to large four-legged mammals. Bell's finding (see chapter 4) that the proportion of children who class whales and spiders as animals decreases from kindergarten through to fourth grade

(1981) probably comes about through children hearing the word applied most frequently to the large land mammals, and rarely to whales, spiders, fish, insects, birds, worms or people. Common usage encourages them to construct a meaning for the word that is different from scientific usage.

In addition to idiosyncratic meanings, we can have different meanings for a word in different contexts. That is, in one context the label encompasses a different set of instances than in another. In chapter 3 I gave the example of signs that say 'No animals allowed on freeway', yet people throng to freeways, including biologists who in another context certainly place humans among the class 'animals'. On returning from abroad, I have to declare to customs officers whether I am bringing in any 'animal products'. In that context I decide that this does not include woollen clothing or my leather briefcase, though I am not sure that I am correct in doing so. If it came to legal action, I could justify scientifically my failure to declare these articles. Communication can fail even when the two people have identical meanings for words if they use and interpret them in different contexts.

Context is especially important in science teaching, because many words have different common and scientific usages: animal, force, energy, weight. Physics teaching is, perhaps, more troubled by this than any other subject. Physics is an abstraction which maps concepts on to the real world, while leaving out many of the details. The trick in interpreting a statement in physics is to know whether details are to be left out, and which ones. The teacher says 'An object will stay in its state of rest or uniform motion in a straight line unless acted on by external force'. Since people do not talk like that in an everyday context, students realize that this is meant to be interpreted in the scientific context. If it were not, it would cause some trouble. In the everyday context, external force would be something deliberately applied, a push or a pull given to the object (a scientific word; 'thing' in everyday language) by a person or some active agency. While gravity might be included as an active agency, friction generally would not. So in the everyday context people accept that things stay still unless pushed, but resist the notion that they will keep going for ever. The unfortunate consequence is that physics is learned not just as an abstraction, but as one with little relevance to the real world.

Even when teachers try to show the application of physics and chemistry, they employ signals to tell the students which context they should be operating in: when you hear *ideal* gas, *smooth* plane, *light* string, *pure* metal, *stiff* rod, you know it is the abstraction. A revealing incident occurred when Gunstone and I were using the prediction–observation–explanation technique in the probe of university students' understanding of gravity that was described in chapter 5. We used the bicycle wheel pulley shown in figure 5.4. I demonstrated that the wheel span freely, and was about to proceed when a student asked, 'Is the wheel frictionless?'. I answered to the effect that nothing is frictionless, but this wheel was well-mounted with good bearings and he had seen it spin without slowing noticeably. 'Yes,' he said, 'but is it meant to be frictionless?' Perhaps more testily than I should, I replied that it was meant to be a bicycle wheel. He had the last word, however: 'Is it *theoretically* frictionless?'. I think the source of this incident lay in our use of common materials in a science room context, so that the student received conflicting signals. The everyday bucket and wheel and block of wood suggested that this was a real context, while their artificial arrangement and the setting suggested that it was to be interpreted as an abstraction. The student was not certain which context we had in mind, and so feared he would not know how to answer the questions we might ask. If we had used line diagrams instead of a real bucket, pulley and block he would have known that the abstract context was meant.

Strangely, little attention has been given in science teaching to the way that failure to judge context correctly leads to misunderstandings. Little attention is paid to the problem in texts on teaching and it is rarely if ever mentioned in teachers' professional magazines. Yet explaining how context affects the meaning of a word could save students from confusion and misunderstanding. While all teachers might help with this explanation, especially teachers of English, science teachers should attend to it in order to bridge the gap between the world of science and the world that students experience outside the classroom.

Sequence of presentation

In addition to constructing sentences well, and considering whether students understand the connectives, qualifiers, nouns and verbs in them, teachers have to arrange the sentences in a sequence that helps the students to construct sensible meanings.

Although it seems obvious that a logical order, in which one statement leads directly to the next, should produce better communication than one in which the statements are jumbled so that each has no relation to those that immediately precede and follow, many studies in the 1950s and 1960s found that random or reverse sequences led to much the same recall and comprehension of texts as logical ones (e.g. Buckland, 1968; Hamilton, 1964; Levin and Baker, 1963; Niedemeyer, Brown and Sulzen, 1969; Payne, Krathwohl and Gordon, 1967). However, in that research work communication was through print, not oral, and the readers may have searched back and forth to create a coherent sequence for themselves from the random arrangement of sentences. Eye movements were not recorded in these studies; presumably logical sequences would provoke fewer vertical eye shifts than would random ones. For oral presentation, Anderson and his co-workers did find that sequence affects the amount of science people learn (Anderson, 1966; Lamb et al., 1979; Mathis and Schrum, 1977; Simmons, 1980). Anderson (1974) proposed that two features of a communication that affect its comprehensibility are the degree of linking between successive sentences and the rate of introduction of new major ideas.

Intuitively, related sentences should have an advantage over disjointed ones. Suppose the teacher says 'Vectors have magnitude, direction, and sense. Money is a scalar'. Both sentences are potentially meaningful, but their apposition causes hearers trouble unless they already know that scalars have magnitude only, and that vectors and scalars are distinct categories of quantities. Without that extra knowledge the sentences are unrelated for the hearers, yet they will try to make sense of them because people are used to the notion that one sentence leads on to the next and illuminates it in some way. We are so conscious of this that we have turns of phrase to indicate when the rule no longer applies – 'To change the subject' or 'By the way'. We do not need phrases to indicate that we are continuing to talk about the same topic. Students know that the stream of information is to be interpreted as a whole, not as separate items. Those few who do not have this cognitive strategy are doomed to failure to understand. The good communicator aids the construction of meaning by presenting sentences in which the logical connection is overt.

As well as the need for successive sentences to be linked, Anderson and co-workers (Anderson and Lee, 1975; Lu, 1978) found that new ideas need to be spaced in their introduction, separated by elaboration of each through examples, explanations and applications. These elaborations assist students to construct meanings for the new notions. Without them students cannot build a meaning for one statement before the next idea is presented. Observation of effective lecturers will show how they follow this principle. Often they will keep a main point before the audience by means of a transparency or writing on a board, and will spend some time talking to that point before revealing the next. The hearers focus on the written point, since that is what the lecturer wants them all to acquire. What they concentrate on or construct from the elaboration may vary.

These two requirements for good structure, of linked sentences and the spaced introduction of new ideas, may seem obvious and uncontroversial, but in practice they are not so easy to maintain. Anderson's analyses found that political speeches were often incoherent through poor linking or erratic introduction of new ideas. President Eisenhower was particularly troubled in this regard. Teachers may suffer from the same faults as politicians. Those who take the trouble to tape some lessons and analyse their linking of sentences and rate of introduction of new ideas, as well as their use of connectives and qualifiers, will discover whether they are making it easy or difficult for their students to construct meaning from what they hear.

Pace of presentation

Although left alone in their classrooms solely in charge of their students, teachers are not completely free to do whatever they like. Apart from legal obligations, there are expectations of the students and their parents, of school authorities and public boards of education that certain things will happen, in particular that a stated amount of subject matter will at least be presented to the students even if that knowledge is not acquired by all. Externally organized examinations are prominent agents in many countries for making these expectations explicit. Whether such examinations are present or not, in most countries schools are organized on an annual basis, with teachers assigned to classes for a year and with a list of topics to be covered in that time. Often the teachers plan what they will do each week by drawing up a program at the start of the year. While this is sensible, it implies that the amount of content to be presented each week is independent of the progress of the students. Of course if they strike difficulty at a point the teacher can slow down, but this must be paid for with hurried treatments of other topics. Teachers are conscious of the external constraint on what they do, and cannot spend the time they might wish on each topic.

Within the bounds permitted by external constraints, judgements of pace depend on teachers' estimates of the states of their classes. Do the students need more help? More elaboration of these points? Have they learned it? Are they interested, or are they getting bored? The teachers' estimates draw on their experiences of other students with that subject matter, from which they form preconceptions of whether the students should find it easy or difficult; also there are the students' responses to questions and the non-verbal signals they display – posture, expression, attention, behaviour. New teachers make more mistakes about pace than experienced ones because only one of these sources of information is readily meaningful to them, the responses to questions. Where an experienced teacher can pick up slight non-verbal signals and decide to change activities, the beginner will often miss them and keep on until the signals become extreme, possibly amounting to classroom revolt. Even experienced teachers have to get used to the reactions and signals of each new class they take, so during the first weeks of every new school year the teacher is learning the signals of the students.

Part of learning how to interpret the reactions of a class is determining who to focus on. From answers to questions and non-verbal signals the teacher forms an interpretation of the class's understanding of the topic, but of course the notion of a 'class's understanding' is a poor one. The class consists of individuals, who may be at all states of comprehension. As familiarity with the class increases, the teacher learns that Anne and Mark always are quick to grasp points, John hardly ever does, and others are in between. Lundgren (1977) found that teachers tend to base their decisions on when to move to new matters on the reactions of a small number of students who are neither the most nor the least able, but whom the teachers regard as being about four-fifths of the way down the ability scale. As far as possible, the teachers do not leave a topic until these marker students appear to comprehend it. The success of a lesson, or sequence of lessons, then depends greatly on the accuracy of the teacher's judgement of comprehension, especially of the marker students. Research on students' understanding of scientific principles, outlined in chapter 5, shows that unfortunately these judgements must often be astray or that external pressures have overborne them. What may be happening is that students are demonstrating ability to perform intellectual skills or that they can recall recently acquired facts, without having much depth of understanding. The shallowness of their comprehension may escape detection by the usual form of classroom questions, while their general level of processing may be so limited that they themselves are satisfied with superficial learning and so transmit non-verbal signals of comprehension and alertness, when greater self-knowledge would have led them to radiate quite different signals of puzzlement or need for more explanation.

The research on understanding indicates that the pace of coverage of science content is too rapid in many countries. On reflection, that is not surprising: curricula compress into a few hours per week for a few years knowledge that has taken some of our most imaginative minds centuries to create. Even though digestion of knowledge is much simpler than its creation, it is probable that we expect too much of students. Research work and the assertion of this book that learning is a time-consuming process requiring reflection on new knowlege and association of it with elements already possessed, imply that decisions on when to proceed should be based on more deliberate and more powerful probes of understanding than are common in current practice. More time should be given to formative assessment.

Useful probes are those described in chapter 5: i.e., concept maps, prediction–observation–explanation tasks, interviews about instances, Venn diagrams and general interviews. Some can be used with a whole class; others, such as general interviews, which take a lot of time and can be run with only one or two students at a time, can be restricted to Lundgren's marker students. Of course extensive use of probes will not become possible unless the syllabus becomes less crowded, teachers are trained to use the techniques and the teachers find that the probes reveal something they did not otherwise know. Only the first of these requirements appears to be a problem, but under present conditions it may be a difficult one to solve.

Questioning

In schools in Western countries, the established script has the teacher asking many questions. Unlike most questions asked outside of schools, which are used to obtain information such as the time or a price or a direction, teachers' questions serve purposes of control: control of behaviour, control of learning and control of the pace of a lesson.

In the previous section teachers' use of marker students (Lundgren, 1977) was described. Control of pace occurs through the responses of those students to questions, which indicate to the teacher whether it is safe to assume that most have grasped a point and it is time to move on or whether more explanation is required.

Behaviour is controlled because the questions focus attention. The students concentrate on what the teacher is doing because they may be called on to answer, and know that failure to respond may be accompanied by sanctions. Many of the questions which, in terms of Bloom's taxonomy of objectives (1956), require knowledge or low-level comprehension, are checks on attention. The teacher fires them quickly round the class and allows the students little time to contemplate them. The teacher's response rewards or punishes the students for their levels of attention.

Attention is the start of learning, as well as part of discipline. Questions direct the students to the events the teacher wants them to focus on, and encourage them to process information at least as far as working memory. Students can hardly remain at either of the lower two levels of attention described at the end of chapter 9 when they know that at any time the teacher may ask them to respond to a simple recall question. However, an even more valuable function of questions is their promotion of deeper processing, the fourth level of attention.

Deep processing occurs when questions force students to link recently acquired information with episodes or older propositions, or when they require them to apply their new knowledge to a problem. An example of the first instance might be, in a lesson on expansion, 'Have you noticed the gaps between railway lines? What happens to them in hot weather?'. Here the teacher is trying to make the students link the proposition that metals expand when heated with an episode of railway lines. Generally, teachers look for a different sort of response to questions intended to encourage linking than to the recall-of-fact type of questions that are used to control behaviour and maintain attention. For the latter, teachers look for definite signals, often of a formal nature, such as a raised hand indicating willingness to answer, and will call on one or more students to respond. For the linking questions, however, they may want only informal signals such as a slight nod or appropriate facial expression. When a quick survey of the informal signals suggests that the linking makes sense to most, or at least to the marker students, the teacher moves on to the next question. However, the model of learning implies that linking is too important to be left to such soft assessment, and that a useful change in teaching style would see more reliable testing of whether links have been made. In class this could consist of the teacher, after presenting the proposition about expansion, asking all the students to write about an example. The writing is important; it is not sufficient to call for oral responses because then only those chosen to answer are sure to have made a link. In tests at a later time new forms of question could be used to check on linking, which would naturally encourage the students to take linking more seriously. One example of a new form is the style of item used by Mackenzie and White (1982) to check whether students linked episodes from an excursion with propositions about the geography of coasts that they had learned in class. An example is shown in figure 10.7. While all of the alternatives are sensible, response C

It is LOW tide and you are standing at the LWM on a mangrove coast. You begin walking back towards HWM.
IT IS DIFFICULT TO WALK—YOU SOMETIMES SINK UP TO YOUR KNEES IN MUD.
Which one of the following facts does this make you think of?

A. Mangrove coasts are spreading seawards.
B. Plants form in zones on a mangrove coast.
C. Soil drainage gets progressively worse across a mangrove coast towards the sea.
D. Tidal range, the difference between HWM and LWM, is large on a mangrove coast.
E. None of these facts.

What else did you think of as you read the situation? (Write on an answer sheet please.)

Figure 10.7 Example of item testing linking of an episode with a proposition (from Mackenzie and White, 1982)

was the one that was looked for as it was the proposition that was emphasized at the time of the experience referred to in the stem of the question. Other forms of item to test links need to be devised.

An example of the second sort of question, that which encourages deeper processing through requiring students to apply knowledge in a problem, is, following lessons on density and flotation, 'Will ice float in methyl alcohol?'. Of course, if the students already know the answer the question is not a problem and fits the knowledge level of Bloom's taxonomy (1956). In framing the question, though, the teacher judges that the students will not have met it before and will have to construct the answer. They would have the proposition 'Solids float in liquids that are denser than they are', and would have to work out that the answer depends on further information: they will have to ask for the densities of ice and methyl alcohol. The chain of reasoning is short, but answering requires thinking about the meaning of the proposition, possibly making many students create images of an ice cube floating, when previously they had not processed the proposition much at all.

Other examples of questions that create problems encouraging deeper processing are:

What limits the length of xylem cells?
Can a weak acid neutralize a strong base?
Why did people choose to make clock hands go clockwise?
If the Earth's orbit were more elliptical, what effects would follow?
Can there be planets round a double star?
What determines the angle of slope of a volcano?
Why is it important to keep seeds of primitive grasses?
Could Superman really stop a speeding train instantly?
What is the evolutionary advantage to dogs in being able to hear higher-pitched sounds than we do?

Answering these questions provokes thought: images are formed, propositions are seen to be related, and episodes of mixing acids and bases, of seeing a sundial, of reading a comic book, of seeing a dog prick up its ears, are recalled. The questions encourage linking of images, propositions and episodes, and so promote extensive patterns of connections in the learners' cognitive structures. Many of these connections will be between topics that the learners had not related previously, such as the structure of plant cells and hydrostatics or the design of clocks and the predominance of the Earth's land masses in the Northern hemisphere.

Although the model of learning predicts that questions that produce more extensive processing will result in better understanding, reviews by Gall (1970) and Winne (1979) indicated that 'higher cognitive questions', like the examples given above, do not bring about better achievement. On the other hand, a later meta-analysis by Redfield and Rousseau (1981) did find a positive effect. The most likely source of the conflict between these results is the dependent variable in the studies. Higher cognitive questions should encourage processing that will be evident in better recall after a long period and greater capability to apply the knowledge to new problems, but should not be expected to have much effect on simple, short-term recall. Indeed, if the task requires verbatim recall, processing can be harmful since it involves representing to oneself the meaning of a communication and the discarding of irrelevant parts of it.

Another reason for conflicting results concerning the value of higher cognitive questions may be breakdown between the asking of the question and the actual processing that occurs. Factors determining breakdown are the difficulty of the question and the time the teacher allows for processing.

Almost by definition, an easy question is one that requires little processing. The answer is found already present in memory. At the other extreme, a very difficult question also may provoke little processing, because the students recognize that they know nothing that is going to help them to answer. Junior secondary students are not going to think hard about questions like 'What difference would it make if the electronic charge were 100 times its actual value?'. For higher cognitive questions to have an effect on understanding and recall they must be pitched at a level that encourages processing. Another difficulty for new teachers is that judging that level is a skill that comes with experience, honed by practice and reflection on the responses obtained from students.

The difficulty of a question depends not only on the students' knowledge, but also on whether they have appropriate cognitive strategies. They have to learn how to answer problem-setting questions. The teacher's role in helping them with that learning involves careful gradation of the problem, encouragement of attempts at answering, and discussion of how to think about them. Trains of reasoning need to be made explicit, so the students can see how problems are solved. They might then generalize the strategy for themselves.

Even if the question is at the appropriate level and the students have the strategies to deal with it, there will not be any processing if the teacher does not allow time for it. Obviously, if the teacher requires an answer or a very quick student supplies one immediately after the question has been asked, then there can be no processing. Time is required to think, so that up to a point delay between question and answer should permit better learning, though if the interval stretched out too long then processing would eventually cease and the students' attention would wander to other things. Rowe (1969, 1974a, 1974b) found that when teachers allowed intervals of three to five seconds instead of the more usual one second, students' responses were longer and they appeared more confident about them; they gave alternative answers rather than being satisfied with one; and they asked more questions back and interacted more with their fellows. Not only were they processing information more deeply, they were taking on more responsibility for the direction of the lesson and for their own learning.

After a student answers the teacher's question there is a second interval, between the answer and the teacher's response. This time is an opportunity for the rest of the class to contemplate the answer and evaluate, accept, reject, qualify or elaborate it. That further processing is also desirable. Tobin (1980) found that longer pauses in this second interval are associated with greater achievement.

Of course a teacher cannot hope to produce dramatic improvements in learning simply by pausing longer. The other two factors of appropriate knowledge and cognitive strategies must be present. If any one of the three is missing, then processing cannot occur. It does appear, though, from the studies of Rowe and Tobin and others that science teachers tend to be too quick in their handling of contemplative questions. As speed is directly under the teacher's control, this is an easy way to increase the probability of better learning.

Although teachers may ask many questions in their lessons, the usual script for classrooms has students asking few. That is unfortunate, because an opportunity to promote understanding is being lost. To form a question involves processing, so most of the thinking in the classroom, like most of the talking, is by the teacher. The outcome is summed up by the trite observation, 'I didn't understand this until I had to teach it'. No doubt having to relate and sequence the various parts of the subject contributes to that conclusion, as does questioning. Barnes (1976) capitalized on the processing induced by teaching by having some students tutor others. The tutored students did about as well as usual, while the tutors showed much better achievement.

Having students ask questions encourages processing. It also gives the teacher new insights about the students' understanding. At present, when students are expected to answer questions but not to frame them, we get some information about what they think but it is inevitably in relation to the teacher's way of looking at the topic. The question limits the responses. A teacher might not think to ask questions like the following, which come from students, but each reveals much about the framer's understanding:

Why doesn't an astronaut fall when he steps out of his spacecraft?
Do people climbing Mt Everest take oxygen so they'll have some gravity?
Do camels with two humps lose one when they go to wetter country?

What is the stuff between atoms made of?
How long does it take the electricity to get from the battery to the globe?

A third advantage of having students ask questions is that it is good training in the cognitive strategy of reflective thinking. Reflection begins with posing questions to oneself. In my values (chapter 1), reflective thinking should be a major objective of science courses. Whatever the propositions and intellectual skills that make up the content of a science course, whether it is chemistry or biology or physics or geology, the students must retain and develop a curiosity about the natural world that is evident in their own questions and reflections.

The skill and habit of forming reflective questions do not occur automatically. They require training and practice. One form of training is to watch others. While students have a model in their teacher, observing an adult is often not as effective as watching someone at their own level of competence. Thus students could learn quickly from each other's public attempts at framing questions. The teacher can assist by constructive comments on each effort, and can reward attempts with praise. There is no reason, either, why tests and examinations should not include items like: 'Write a question beginning with "What if . . ." about density'. Students would see that question-asking is taken seriously by the teacher.

Suchman (1966) devised a specific technique for training students to frame questions in science. The teacher demonstrates a puzzling, perhaps paradoxical situation; for example, two identical pots, one containing a healthy plant and the other a dead one, under glass jars. The students have to form an explanation by asking questions that must be framed so that the teacher can answer them with either a 'yes' or a 'no'. They can confer, and can stop when they are satisfied with a hypothesis (say, that although the pots, plants, soil, water and light were identical, one jar contained an atmosphere of nitrogen only). The teacher does not reveal the answer, but comments on the questions and helps the students improve their skills of framing them.

Although the model of learning advocates that students should ask questions, and practical procedures exist for training them to do so, there are obstacles to the practice becoming widespread. Teachers find it hard to separate control of behaviour from control of the course of a lesson, and treat the rare questions from students as attempts to sidetrack the direction from the main line along which the teacher is determined it shall go. Even though a student's question may be treated courteously, the usual practice is for it to be dealt with briefly and then left in order to return to the predetermined course. There is no extensive following-up of the student's interest. At least in part this is because the teacher has a fixed syllabus to cover in a limited time. Also, to follow one student's interest may not suit the rest of the class. They might not want to follow the teacher's line either, but do so because that is an accepted part of the script.

Although obstacles exist, students can be trained to ask questions. That was one aim of the large action research project, PEEL, at Laverton High School on the outskirts of Melbourne. Training employed in the project enabled students to make considerable progress in the number and quality of questions that they asked (Baird and Mitchell, 1986). This should have a beneficial effect not only on the quality of their learning but also on their interaction with the world after school.

The laboratory

The laboratory sets science apart from most school subjects. It gives science teaching a special character, providing for many teachers and their students liveliness and fun that are hard to obtain in other ways. That character is almost sufficient alone to justify the high capital and recurrent costs of laboratories. Many students are attracted to science, not because of the insights it provides about the natural universe but because of the colour, mystery and oddness of the equipment and materials that are present in the laboratory. Many like the purity of chemicals and their appearance and odour, the intricacy of physics apparatus, the strangeness of the world under a microscope. Science teachers who never outgrow that attraction are fortunate indeed, and so are their students for that enthusiasm will be communicated. However, it would be even better if the laboratory were arranged to serve other purposes as well; our model of learning indicates what these purposes might be, and how they might be achieved.

Obviously the laboratory provides training in motor skills. It could also be the source of episodes and images that give meaning to propositions that students have already learned or will acquire. The laboratory can provide training in the cognitive strategies of problem-solving and learning. It may be an efficient way of communicating understanding of, and the skills associated with, scientific method. Without the laboratory, it could be difficult for students to comprehend what scientists do.

Values and beliefs about learning determine whether a teacher wants the laboratory to promote many objectives equally or to emphasize only a few. Certainly people differ in opinions about the purpose of laboratory work. From their particular view of learning theory, Kreitler and Kreitler (1974) argued that laboratory experiences help students establish the accuracy of beliefs. They must therefore promote reflection on knowledge and the resolution of conflicting principles, which, as we saw in chapters 5 and 6, are serious deficiencies in current learning. The laboratory, for Kreitler and Kreitler, also provides direct experience with concepts, that is, leads to episodes that give the concepts meaning. They do not agree that major functions include training in problem-solving or arousing curiosity. Surveys uncover different appreciations. Gould (1978) found that teachers had shifted over a 20-year period away from the laboratory as an aid to understanding towards a view of it as a source of interest and as a training ground in problem-solving. In our model, training in problem-solving and understanding are likely to go hand-in-hand, but Gould's teachers and Kreitler and Kreitler apparently could distinguish them as unconnected processes. Students may see the use of the laboratory as different again. Osborne (1976) found that first year undergraduates thought that the laboratory was more effective than tutorials, lectures and other settings for creating interest in physics and for developing critical thinking, but the worst method for training in problem-solving. A similar result was obtained in a survey by Boud et al. (1980). In contrast, Ben-Zvi et al. (1976) found that tenth-grade students in Israel placed promotion of interest in chemistry in last place for a set of eight aims.

Since the opinions expressed by theorists and respondents to surveys are affected by their values and experiences, it should not be surprising that they differ in their ratings of the various purposes of laboratory work. Whatever one's view, what matters is whether the laboratory actually fulfils its function. Early research (Kruglak, 1953; Yager, Engen and Snider, 1969) suggested that it does not, except for manipulative skill. Certainly the surveys reported by Osborne (1976) and Boud et al. (1980) indicate that students think they learn little from it about problem-solving. Better manipulative skill hardly seems sufficient return for the resources committed to laboratories, yet despite the discouraging research results, faith in the laboratory persists.

Faith in the laboratory may be justified, even though current practice in it is not. The laboratory may not have had a good effect on interest or understanding or problem-solving because it was inappropriate laboratory. There is evidence that students often have no clear purpose in the laboratory, and can therefore learn little. Moreira (1980) asked undergraduates five questions, to probe their understanding of purpose and their involvement, after they had done each of four experiments. The questions asked for the problem the experiment was about, the relevant key concepts, the phenomena involved, the method followed, and the answer obtained. The responses indicated that few students understood what they were doing. When Tasker (1981) talked with 11- to 14-year old students during laboratory sessions, he found that they usually perceived the sessions as isolated events, not related to other lessons, and that they often could not state the purpose of the experiment or describe the place of each part in it:

Observer: What have you decided it [the task] is about?
Pupil: I dunno, I never really thought about it . . . just doing it – doing what it says . . . it's 8.5 . . . just got to do different numbers and the next one we have to do is this (points in text to 8.6).

(Tasker, 1981, p. 34)

Clearly, these students' learning was largely finding out how to follow directions.

Our model of learning suggests that as well as providing training in motor skills, the laboratory should be a source of episodes that form attitudes, make propositions meaningful and develop cognitive strategies useful in learning and problem-solving. The early research results, and Moreira's and Tasker's observations, indicate that this happens rarely. What can be done to make the laboratory effective? We need to think about the sorts of experience arranged to occur there, and the details of procedure.

The laboratory is a source of episodes, which range from the unique to the commonplace. Wherever their place in that range, they can be valuable. The unique ones, that remain in memory as 'unforgettable' events, can be pillars to which are anchored propositions and intellectual skills, while the common ones merge into scripts that make concepts understandable.

In current practice, laboratories are planned to be orderly, smoothly organized places without dramatic events. Anything unusual is an accident, perhaps even a disaster, that does not help learning – though I recall an episode from my school days in which explosions caused burns and cuts, an episode that is linked for me with the proposition that ammonium nitrate decomposes with heat, but it is a memory acquired at doubtful cost. Instead of wasting the laboratory's potential for drama on valueless accidents, it would be sensible to place in a year's program a few experiments that are unusual enough to establish specific episodes, to which important knowledge can be linked. Their number must be small, because drama every week becomes humdrum. Therefore the topics they anchor must be chosen carefully. In physics, for example, transformation and conservation of energy are key principles, so it would be useful to have a remarkable energy experiment. Perhaps, with due attention to safety, students could make a powerful catapult, and calculate energy values and changes when it is stretched and fired.

The teacher has to do more than arrange a spectacular event. The students may form episodes, but that will have no great value if they do not link those episodes to other information. Teachers have to promote that linking, and may invent more imaginative ways of doing so than the two suggestions that follow. First, linking would be enhanced by the students creating their own explanations of the purposes of the experiment. Someone who does not know the purpose will find it difficult to link the episode with propositions. Tasker's and Moreira's observations show that it is not sufficient for the teacher to write out the aim, because students often fail to process it. Nor is it sufficient to have students write a single bland sentence such as, 'The aim is to study transformation of energy in a catapult'. Rather, they should be trained to write clear paragraphs, something like this say, for upper secondary students:

Our aim was to construct a catapult and make measurements of the energy stored in it and released when we fired it. We were to compare the stored and released quantities, and if the former was greater as expected, to suggest where the missing energy had gone to.

They must, of course, compose these paragraphs themselves, otherwise there is little or no processing. Learning how to write them requires guidance at first, and helpful criticism.

The second suggestion is that, at the end of the experiment, the students should list all the propositional knowledge they used or thought of in the experiment. The list might be:

Catapults were used in ancient wars.
Force is a push or pull.
Hooke's Law: extension proportional to force.
Energy is area under force–extension graph.
Rubber gets warm when stretched lots of times.
Hysteresis has something to do with energy loss.
Energy is conserved.
A stretched rubber band has potential energy.
If Hooke's Law holds, energy $= \frac{1}{2}kx^2$.
The greater the velocity the greater the range.
Air resistance reduces range.
Spinning objects have energy.
Moving things have kinetic energy.
The angle for maximum range is 45°.

Constructing such lists should be more valuable for students than the calculations and details of procedure that make up current laboratory reports. Comparison and discussion of lists would increase the learning they have already promoted.

Spectacular experiments are peaks above the plain of common ones, which also have a function. While the everyday experiments do not lead to specific episodes, they enable students to form scripts and images that give meaning to propositions and skills. Repeated experiences with chemicals give meaning to words like 'reaction', 'precipitate', 'solution', so that sentences like 'magnesium reacts with hydrochloric acid to evolve hydrogen' and 'ferric hydroxide is precipitated when sodium hydroxide is added to a solution of a ferric salt' are readily processed. Without experience some meaning might be derived from statements like those, but it is easier when images can be drawn from generalized experiences. Laboratories do provide experiences that are useful in this way, but we should ask whether the experiences can be even better.

One weakness of laboratory episodes is their 'labness', their lack of relation to the materials and experiences that students encounter out of school. This means that school-acquired knowledge remains apart from everyday matters, so that it is difficult to meet the aim for science to illuminate people's lives. If a substantial proportion of laboratory investigations used common materials instead of things never encountered elsewhere, the gap would be bridged. While this would be difficult or impossible in some topics, many others provide opportunities. Kinematics experiments can be done with people, bicycles and cars instead of trolleys and airtracks; refractions can be seen and measured in swimming pools as well as in glass blocks; electricity can be studied with house fuses, switches and meters rather than with rheostats and potentiometers; acids can be found in fruit and bases in household cleaners as well as in glass jars. The episodes formed during experiments with common materials would link science propositions and intellectual skills to the students' broader knowledge.

The procedures mentioned above of describing the purpose of experiments and listing related knowledge can be used with undramatic as well as more spectacular experiments, and with those using common materials as well as 'laboratory' equipment. A less direct but powerful way of promoting inter-linking of knowledge is to have the students apply it to problems, which has the further benefit of providing training in cognitive strategies. Unfortunately, current laboratory practice rarely includes real problems. Most experiments are exercises, providing little more than acquaintance with materials and objects and training in following directions. Probably the reason for this is the pressure of time. Laboratory sessions have to fit into a weekly schedule, and a certain quantity of material has to be presented, if not learned well, each lesson. Real problems take time, and the amount of time can hardly be specified and planned for. Yet their value is surely sufficient for some such problems to be included in the curriculum each year. A teacher I met in England gives her students problems that link real social issues with physics – construct a door bell and an alarm clock for the deaf, an indicator that blind people can use to tell how full a saucepan is of a hot or dangerous liquid, an alarm system to tell whether a baby has been taken from a pram. Other examples of real problems are measuring the mass of a car, the potential energy in an inflated balloon, the strengths of adhesives, the speeds of reactions of animals.

To sum up, laboratory experiences should be a mix of dramatic and ordinary experiments, and of directed and problem-solving activities, using exotic and everyday materials. Linking of laboratory episodes with propositions and intellectual skills should be promoted through the students writing extended descriptions of the purpose of each experiment and lists of the knowledge to which the experiment is related.

Style, and three principles of teaching

Although one script fits nearly all classrooms, teachers do vary from each other, and even vary in each one's own behaviour from time to time in emotional warmth, number of words spoken per lesson, types of questions used, proportions of time they keep their pupils on task, and in many other dimensions that make up their styles. Of the many dimensions, I want to consider briefly the form of direction that the teacher uses to control the learning of the class, because it leads to three principles of teaching that relate to the theme of this book that learning involves the processing of information and the construction of meaning for it.

While style of direction can be described in many ways, a convenient system for science education is that devised by Galton and Eggleston (1979): teachers as informers, problem-solvers or inquirers. The names of the categories may be sufficient description of the styles.

In their observation Galton and Eggleston found that the teachers rarely changed their styles: once an informer, always an informer, seemed to be the rule. Theobald (1980) also found consistent behaviour when he observed biology teachers on five occasions each over a year. Of course teachers are not automatons and do vary their styles to a degree, perhaps when dealing with different content or different combinations of personalities in classes of students. However, the research and one's own experience suggest that teachers evolve a style and change its general form slowly, over years, if at all. Therefore it is not too distorting to accept the categories of Galton and Eggleston as representing consistent styles. The categories are stereotypes, of course, representing extremes, with each individual having a mean position somewhere in the space between them and oscillating a little from that personal mean at each lesson.

Whatever the teacher's style, the students have to cope with it. With an archetypical informer, they are going to have to process sentences, and can be expected to develop strategies for doing so. Their knowledge will consist largely of propositions, which will not be linked with episodes. They might acquire strategies of linking one proposition with another, though that will depend on their experiences outside school, such as the things they read or that their parents and friends talk about, and on the range of topics the teacher informs them about. For within the informing style there will be variations, from the teacher who sticks closely to the matter in hand, treating each topic as a packet of knowledge complete in itself, to the enthusiast who sees connections between this topic and many others and tries to alert the students to them.

The outcome of learning from an informing teacher will depend also on the pace of presentation. Rapid delivery of information inhibits processing, so that only a small proportion of sentences will be stored as meaningful propositions. Slower delivery allows students to contemplate the meaning of each sentence and its relation to others.

Variations in diversity of topics and pace of delivery, along with other differences to do with personality and the structure of the communication, will produce differences in the quality of the learning that occurs. They will influence the formation of cognitive strategies. However, variations within the informing style will not encourage formation of those strategies that are an integral part of finding out knowledge for oneself. Long exposure to informing teachers, whether they are good or bad, brings people to see learning as something that is done to them, not something that they control. It is hard to see how people who have had this experience will develop the following strategies defined by Baird (1984): evaluating degree of understanding to decide whether further learning is required, deciding what should be learned next, considering purpose of the learning, reflecting on significance of information to one's own life, evaluating own readiness to learn. Failure to develop them will, in turn, affect the structure of the learner's knowledge.

In similar ways there can be variations of a better or a worse nature within the other two archetypes – problem-solvers and inquirers – which again will mould the pattern of cognitive strategies that the student develops.

Although there are better teachers and worse ones, there is no one style that all should be encouraged and trained to follow. Each style tends to bring about certain outcomes, but the relation is blurred by the many variations in other dimensions that were not used in describing the style. The relation is affected also by the students' familiarity with the style. A superb practitioner of the inquirer type will have a different experience with students who are not used to that style than with those who know how to benefit from it, and quite different learning will occur in those two sets of students.

This discursion on style provides the background for stating three principles of teaching – the principle of maximum opportunity for processing, the principle of matching teaching style with learning style and both with content, and the principle of balance.

Principle of maximum opportunity

Whatever the style and personality of the teacher, and whatever the capabilities of the students, variations in pace of presentation will affect the amount of processing or construction of meaning that occurs. The relation is of the form shown in figure 10.8, an inverted U. If the pace is too rapid, the learners are swamped with information and little will get past the short-term store. As the pace increases even further, a point is reached where there is not even translation. At the other extreme the curve also dips because although there is time to process each new piece of information, there is only a limited amount that can be done to it, and when that has been done processing will stop. Lessons in which only one fact is presented or one intellectual skill acquired, which are more common than lay people might imagine, can involve little processing.

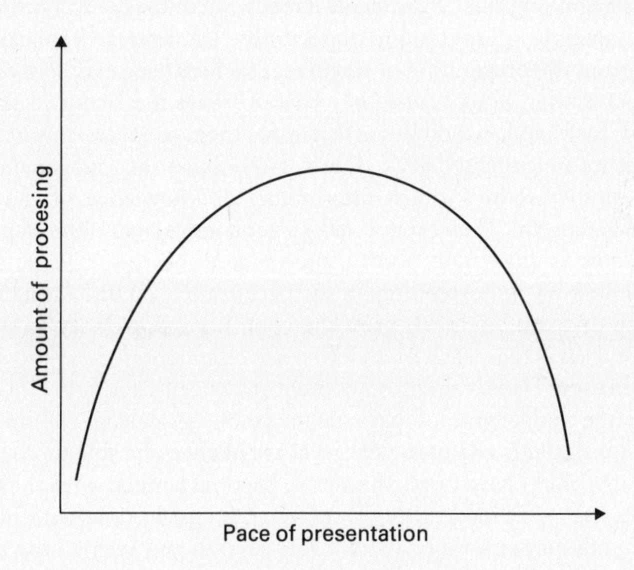

Figure 10.8 Postulated form of relation between pace of presentation and amount of processing

It would be convenient if the optimum rate, allowing the greatest amount of learning, could be determined and a simple recommendation given, such as 'introduce six facts per hour'. Unfortunately the rate depends on too many factors: the newness of the information to the learners, their abilities, their physical state, their interest in the topic, the style of the teacher – essentially the factors shown in figure 2.1. Their interplay changes the scales on figure 10.8 so that, in a horizontal shift of the curve, under one set of conditions the students can benefit from a rapid flow of information while under others they need a lot of time to construct meaning for an item; and, vertically, the maximum amount of processing that is possible may sometimes be great and sometimes small.

One of the chief skills in teaching is judging accurately how much processing is possible and what the pace of presentation must be to allow students to attain that maximum. What makes teaching all the more a matter of skill is that the students in a class differ in both the pace at which their maximum occurs and its height, and so either a means must be adopted to allow each to attain the personal maximum, or, more commonly, a common pace has to be selected in order to have the most tolerable result overall – not so fast that some are lost and not so slow that some are unstimulated. This is where teachers make use of the marker children that Lundgren (1977) identified. Learning these judgements is a matter of experience, and the lack of this ability is one factor that makes teachers beginning their careers less effective than they later become.

The principle of maximum opportunity has several implications for teaching and the preparation of teachers. Teachers need excellent command of subject matter, not because they have to share the facts and skills with students, for then it would be sufficient for them to know only as much as the students are going to have to know, but because their judgement of pace involves knowing how much interlinking of knowledge of the topic is possible and useful. They cannot make a good judgement if their knowledge of the topic is slight and poorly inter-related. Nor, of course, can they promote linking and extending on the part of their pupils. An unfortunate inference that can be drawn from the continuing popularity in science of formalist texts is that teachers are often not confident in their knowledge of subject matter. Their training on content may need attention.

Since the judgement of pace depends on knowledge of how much processing students at a particular level are likely to be able to do, teachers in training should have opportunities to become familiar with the thinking of students. A sensible practice is for them to spend time with individual children, talking with them about a fact or skill and seeing how the child links it to other knowledge.

The principle of maximum opportunity implies that class teaching, where all must keep at the same rate, is less efficient than procedures in which each student can work at a personal optimum rate. In real practice, few teachers use a free-rate method because of constraints of the syllabus and the difficulty of keeping track of what students are doing and supplying each with useful individual attention. The constraints and difficulties may be less severe than many teachers imagine, but certainly appear difficult enough to dissuade most from trying.

Personal optimum rate depends in a complex way on the cognitive strategies one can apply. A person with no strategies for elaborating information is a poor learner, who does not benefit from increased time. That person's optimum rate is quite fast. Oddly, so is the rate for someone near the other extreme who has useful strategies under full command and can apply them speedily. In between are people who need time to reflect and so to learn well. Then, beyond the able, speedy processors, there are exceptional learners who again have a slower optimum rate because, although they have good strategies, they engage in reflection so deep that the complicated patterns of their thought simply take a long time. These are the able students observed by Krutetskii (1976), who took longer over mathematics problems than those slightly less able because they were thinking about the general class to which each problem could belong. That is, they were linking it to other intellectual skills.

Principle of matching

Discovery learning has long been advocated in science teaching. The model of learning provides theoretical support for it, since discovery must involve deep processing and construction of meaning – though that is not to say that construction of meaning is not involved in reception of information, too. The model also suggests that it is over-simple to think that discovery can be implemented directly in any situation. One of the model's principles is that learning involves application of cognitive strategies that are not innate but are built up over an extended period from experience, reflection and guidance. Although they are general skills, applicable to wide ranges of tasks, some strategies will be required for discovery learning which are not learned from experience of didactic teaching. Therefore sudden introduction of discovery methods to students who have no previous acquaintance with them will cause confusion rather than an immediate gain in performance.

Any sudden innovation in education will most likely be followed by a decline in desired outcomes until all involved learn new strategies. I experienced this in Thailand when, as consultant to a large and progressive curriculum project, I was asked to help train teachers to ask questions. Until then, the standard script had the teacher talking almost continuously, with students' responses restricted to answering in unison simple questions that required no more than recall of a string or a well-drilled fact. The aim of the project was to introduce discovery learning, so it seemed a sensible step for the teachers to learn how to ask questions that encouraged students to reflect on their knowledge. In order to make a videotape for use in in-service courses, I rehearsed a teacher in the asking of questions, and together we arranged for him to be recorded teaching one of his usual classes. Our first attempt failed because, although he asked the questions well, the students had no script for the new situation and no strategies for constructing meaning from it or for forming answers. They sat mute. As with all hindsight, it was obvious afterwards what had gone wrong. The Thais were then able to introduce the innovation more sensibly than I had tried to do, by attending to the preparation of students for the curriculum change as well as the teachers. It is interesting to reflect that no consideration was given to training students for curriculum innovation when PSSC, Chem Study, BSCS, Nuffield, ASEP or other major courses were introduced in the United States, Britain or Australia.

My anecdote about Thailand is an illustration of the general principle that the teaching method and the students' learning strategies must match for immediate processing of new information. The second part of the principle is less obvious, that for the long-term development of new strategies, method and strategies must *not* match exactly. If there is no strain between the teaching and the strategies, learning can proceed easily but there is no demand on the students to acquire new ways of learning. They will merely be reinforced in the strategies they have always found to work. That does not matter if the situation mirrors fully the life they will have outside school. However, although there is much receptive learning in general life, with people obtaining information from television, newspapers, books and conversations, there is also the opportunity to learn a great deal from observation, manipulation of materials and reflection on experience, which in chapter 1 was rated high as a value for science education. It is desirable, then, for students to acquire a broad range of strategies, which they can apply in diverse situations. If they are to acquire such a range, learning should not be made too smooth and comfortable a process for them.

A third consideration in matching is the type of outcome. The teaching and learning styles have to be fitted to the nature of what is to be learned. When strings are being acquired, didactic teaching combines well with students' ability to concentrate on the task. When Socratic questioning is added to the teaching, strategies for reflection on knowledge are required. Outcomes such as relating propositions to each other are then attainable, although others, such as skills of designing scientific investigations, cannot be learned in this way.

The interplay of two of the factors in this system, style of teaching and type of outcome, is widely recognized as important even if the fine details of it are often not understood. However, the role of the other factor, the learning strategies of the students, is not greatly appreciated. Educationists often make the mistake that I did in Thailand, of thinking that a change in outcomes can be brought about directly by a change in teaching style alone. Instances in the learning of science where this observation is especially apt are laboratory work and excursions. Students come to these situations from other lessons in science and lessons in almost all their other subjects in which they are used to applying strategies appropriate for reception learning. As the studies by Tasker (1981), Moreira (1980) and Mackenzie and White (1982) already mentioned have shown, students often do not benefit from the laboratory or excursion experience, probably because they do not know how to. Science teachers should thus first train their students in strategies of learning from laboratory work or a field trip.

Principle of balance

Several points from the descriptions of the two principles already discussed are reiterated in the third principle, which also takes in other matters from the model of learning. This principle is that the practice of teaching involves keeping many things in dynamic balance.

Teaching should not be totally didactic, for then students would not learn to construct meaning for the information; nor can it be based entirely on discovery, because that would be inefficient, with not enough knowledge being acquired coherently for much to be understood. It is not good to employ a single teaching style, because students will then develop only a restricted set of cognitive strategies and may never acquire some that are essential in other situations; but too rapid an alternation of styles inhibits acquisition of strategies as the students never have long enough experience with any one. All control should not be in the hands of the teacher, because the students would not see learning as an act for which they are responsible and would become indifferent to the information given to them; but too much control cannot be given to students because they do not know enough about what they do not know to choose the best direction to follow.

The principle of balance asserts that excess in any direction in the arrangement of conditions of learning is bound to bring penalties. The problem for the teacher is to recognize what constitutes excess. The median of social activities is a mirage, hard to fix, and so the direction in which to move to maintain balance is not easy to determine. The directions in which people want things to move depend on their values and their perception of how things are, that is, on their construction of the meaning of the situation. Thus some may see schools as repressive and constricting places that need to move towards freedom and fewer rules, while others see them as blots of anarchistic chaos that need greater direction and clarity of purpose. Until everyone has the same values and perceptions, and that is to say never, such conflicts will persist.

Difficulty in identifying the centre of balance and the shifts necessary to approach it do not negate the principle, nor does the recognition that there are many dimensions on which balance or imbalance can occur. The image is one of balancing a spinning disc rather than a stationary rod. Teachers are jugglers who apply principles that they hone with practice. Principles alone do not make an effective teacher, but nor does practice, for not all experienced teachers are good teachers. To benefit from experience teachers need principles with which they can compare their practices. The principles of learning described in this book are intended to be tools that teachers can use to help their students to learn science well.

References

Allport, G. W. 1935: In C. Murchison (ed.), *A Handbook of Social Psychology*. Worcester, Mass.: Clark University Press.

Anderson, O. R. 1966: The strength and order of responses in a sequence as related to the degree of structure in stimuli. *Journal of Research in Science Teaching*, 4, 192–8.

Anderson, O. R. 1974: Research on structure in teaching. *Journal of Research in Science Teaching*, 11, 219–30.

Anderson, O. R. and Lee, M. T. 1975: Structure in science communications and student recall of knowledge. *Science Education*, 59, 127–38.

Andersson, B. and Kärrqvist, C. 1979: *Elektriska Kretsar*. Götheborg, Sweden: Department of Educational Research, University of Götheborg, Elevperspectiv 2.

André, M. E. D. A. and Anderson, T. H. 1978–9: The development and evaluation of a self-questioning study technique. *Reading Research Quarterly*, 14, 605–23.

Archenhold, W. F., Driver, R. H., Orton, A. and Wood-Robinson, C. 1980: *Cognitive Development in Science and Mathematics*. Leeds: University of Leeds.

Aristotle 1956 (wr. 4th century BC): *Parts of Animals* (A. L. Peck, trans.), London: Heinemann.

Ausubel, D. P. 1963: *The Psychology of Meaningful Verbal Learning*. New York: Grune and Stratton.

Ausubel, D. P. 1968: *Educational Psychology: A Cognitive View*. New York: Holt, Reinhart and Winston.

Bagozzi, R. P. and Burnkrant, R. E. 1979: Attitude organization and the attitude–behavior relationship. *Journal of Personality and Social Psychology*, 37, 913–29.

Baird, J. R. 1984: 'Improving Learning through Enhanced Metacognition.' Unpublished PhD thesis, Monash University, Melbourne.

Baird, J. R. 1986: Improving learning through enhanced metacognition: a classroom study. *European Journal of Science Education*, 8, 263–82.

Baird, J. R. and Mitchell, I. J. 1986: *Improving the Quality of Teaching and Learning: An Australian Case Study - The PEEL Project*. Melbourne: Monash University.

Baird, J. R. and White. R. T. 1982a: A case study of learning styles in biology. *European Journal of Science Education*, 4, 325–37.

Baird, J. R. and White, R. T. 1982b: Promoting self-control of learning. *Instructional Science*, 11, 227–47.

Baird, J. R. and White, R. T. 1984: 'Improving Learning through Enhanced Metacognition: A Classroom Study.' Paper presented at the meeting of the American Educational Research Association, New Orleans.

Barnes, D. R. 1976: *From Communication to Curriculum*. Harmondsworth: Penguin Books.

Bell, B. F. 1981: When is an animal, not an animal. *Journal of Biological Education*, 15, 213–18.

Bell, B. F. 1984: 'The Role of Existing Knowledge in Reading Comprehension and Conceptual Change in Science Education.' Unpublished doctoral dissertation, University of Waikato, Hamilton, NZ.

Ben-Zvi, R., Hofstein, A., Samuel, D. and Kempa, R. F. 1976: The attitude of high school students towards the use of filmed experiments. *Journal of Chemical Education*, 53, 575–7.

Berman, L. M. 1968: *New Priorities in the Curriculum*. Columbus, Ohio: Merrill.

Bloom, B. S. (ed.) 1956: *Taxonomy of Educational Objectives: Handbook 1. Cognitive Domain*. New York: Longmans, Green.

Bloom, B. S. 1968: Learning for mastery. *Evaluation Comment*. Los Angeles: Center for the Study of Evaluation of Instructional Programs, UCLA.

Blumer, H. 1955: Attitudes and the social act. *Social Problems*, 3, 59–64.

Blythe, R. 1969: *Akenfield: Portrait of an English Village*. London: Allen Lane, the Penguin Press.

Bogardus, E. S. 1925: Measuring social distances. *Journal of Applied Sociology*, 9, 299–308.

Boud, D. J., Dunn, J., Kennedy, T. and Thorley, R. 1980: The aims of science laboratory courses: a survey of students, graduates and practising scientists. *European Journal of Science Education*, 2, 415–28.

Broadbent, D. E. 1975: The magic number seven after fifteen years. In A. Kennedy and A. Wilkes (eds), *Studies in Longterm Memory*. London: Wiley.

Brooks, A. 1986: (untitled chapter) In J. R. Baird and I. J. Mitchell (eds), *Improving the Quality of Teaching and Learning: An Australian Case Study - The PEEL Project*. Melbourne: Monash University.

Brown, A. L. 1980: Metacognitive development and reading. In R. J. Spiro, B. C. Bruce and W. F. Brewer (eds), *Theoretical Issues in Reading Comprehension: Perspectives from Cognitive Psychology, Linguistics, Artificial Intelligence and Education*. Hillsdale, NJ: Erlbaum.

Brown, A. L and Smiley, S. S. 1977: Rating the importance of structural units of prose passsages: a problem of metacognitive development. *Child Development*, 48, 1–8.

Brumby, M. 1979: Problems in learning the concept of natural selection. *Journal of Biological Education*, 13, 119–22.

Bruner, J. S. 1966: *Toward a Theory of Instruction*. Cambridge, Mass: Harvard University Press.

Buckland, P. R. 1968: The ordering of frames in a linear program. *Programmed Learning and Educational Technology*, 5, 197–205.

Buros, O. K. 1953: *The Fourth Mental Measurements Yearbook*. Highland Park, NJ: Gryphon Press.

Burt, C. 1966: The genetic determination of differences in intelligence: a study of monozygotic twins reared together and apart. *British Journal of Psychology*, 57, 137–53.

Campbell, J. R. 1972: Is scientific curiosity a viable outcome in today's secondary school science program? *School Science and Mathematics*, 72, 139–47.

Chambers, D. W. 1983: Stereotypic images of the scientist: the draw-a-scientist test. *Science Education*, 67, 255–65.

Chambers, W. and Chambers, R. 1869: *Class-book of Science and Literature*. London and Edinburgh: Chambers.

Champagne, A. B., Klopfer, L. E. and Anderson, J. H. 1980: Factors influencing the learning of classical mechanics. *American Journal of Physics*, 48, 1074–9.

Champagne, A. B., Klopfer, L. E., DeSena, A. T. and Squires, D. A. 1981: Structural representations of students' knowledge before and after science instruction. *Journal of Research in Science Teaching*, 18, 97–111.

Chase, W. G. and Simon, H. A. 1973: The mind's eye in chess. In W. G. Chase (ed.), *Visual Information Processing*. New York: Academic Press.

Chesterton, G. K. 1929: *The Father Brown Stories*. London: Cassell.

Comte, A. 1855: *The Positive Philosophy* (H. Martineau, trans.), New York: Blanchard. (Original work published 1830–42; this edition reprinted New York: AMS Press, 1974.)

Cook, L. K. 1982: 'The Effects of Text Structure on the Comprehension of Scientific Prose.' Unpublished doctoral dissertation, University of California, Santa Barbara.

Dansereau, D. F., Brooks, L. W., Holley, C. D. and Collins, K. W. 1983: Learning strategies training: Effects of sequencing. *Journal of Experimental Education*, 51, 102–8.

Dibley, L. 1986: (untitled chapter) In J. R. Baird and I. J. Mitchell (eds), *Improving the Quality of Teaching and Learning: an Australian Case Study - the PEEL Project*. Melbourne: Monash University.

Dick, O. L. 1949: *Aubrey's Brief Lives*. London: Martin Secker and Warburg.

Doob, L. W. 1947: The behavior of attitudes. *Psychological Review*, 54, 135–6.

Driver, R. 1983: *The Pupil as Scientist?* Milton Keynes: Open University Press.

Driver, R., Guesne, E. and Tiberghien, A. (eds) 1985: *Children's Ideas in Science*. Milton Keynes: Open University Press.

Duit, R., Jung, W. and von Rhöneck, C. (eds) 1985: *Aspects of Understanding Electricity: Proceedings of an International Workshop*. Kiel: Institut für die Pädogogik der Naturwissenschaften an der Universität Kiel.

Dunkin, M. J. and Biddle, B. J. 1974: *The Study of Teaching*. New York: Holt, Rinehart and Winston.

Eisner, E. W. and Vallance, E. 1974: Five conceptions of curriculum: Their roots and implications for curriculum planning. In E. W. Eisner and E. Vallance (eds), *Conflicting Conceptions of Curriculum*. Berkeley, Ca: McCutchan.

Erickson, G. L. 1979: Children's conceptions of heat and temperature. *Science Education*, 63, 221–30.

Erickson, G. L. 1980: Children's viewpoints of heat: a second look. *Science Education*, 64, 323–36.

Fensham, P. J. 1984: Conceptions, misconceptions, and alternative frameworks in chemical education. *Chemical Society Reviews*, 13, 199–217.

Fishbein, M. and Ajzen, I. 1974: Attitudes towards objects as predictors of single and multiple behavioral criteria. *Psychological Review*, 81, 59–74.

Flanders, N. A. 1970: *Analyzing Teaching Behavior*. Reading, Mass.: Addison-Wesley.

Flavell, J. H. 1976: Metacognitive aspects of problem solving. In L. B. Resnick (ed.), *The Nature of Intelligence*. Hillsdale, NJ: Erlbaum.

Fraser, B. J. 1975: The impact of ASEP on pupil learning and classroom climate. *Research in Science Education*, 5, 1–12.

Fraser B. J. 1980: Science teacher characteristics and student attitudinal outcomes. *School Science and Mathematics*, 80, 300–8.

Fredette, N. and Lochhead, J. 1980: Student conceptions of simple circuits. *The Physics Teacher*, 18, 194–8.

Gagné, R. M. 1962: The acquisition of knowledge. *Psychological Review*, 69, 355–65.

Gagné, R. M. 1965: *The Conditions of Learning*. New York: Holt, Rinehart and Winston.

Gagné, R. M. 1968: Learning hierarchies. *Educational Psychologist*, 6, 2–9.

Gagné, R. M. 1972: Domains of learning. *Interchange*, 3, 1–8.

Gagné, R. M. 1977: *The Conditions of Learning*, 3rd edn. New York: Holt, Rinehart and Winston.

Gagné, R. M. and White, R. T. 1978: Memory structures and learning outcomes. *Review of Educational Research*, 48, 187–222.

Gall, M. D. 1970: The use of questions in teaching. *Review of Educational Research*, 40, 707–21.

Galton, M. and Eggleston, J. 1979: Some characteristics of effective science teaching. *European Journal of Science Education*, 1, 75–86.

Gardner, P. L. 1974: Changes in attitudes of PSSC Physics students: a third look. *Australian Science Teachers Journal*, 20(1), 99–104.

Gardner, P. L. 1977a: Logical connectives in science: a summary of the findings. *Research in Science Education*, 7, 9–24.

Gardner, P. L. 1977b: *Logical Connectives in Science*. Melbourne: Monash University.

Garskof, B. E. and Houston, J. P. 1963: Measurement of verbal relatedness: an idiographic approach. *Psychological Review*, 70, 277–88.

Gauld, C. 1986: Models, meters and memory. *Research in Science Education*, 16, 49–54.

Geeslin, W. E. and Shavelson, R. J. 1975: An exploratory analysis of the representation of a mathematical structure in students' cognitive structures. *American Educational Research Journal*, 12, 21–39.

Ghatala, E. S., Levin, J. R., Pressley, M. and Lodico, M. G. 1985: Training cognitive strategy-monitoring in children. *American Educational Research Journal*, 22, 199–215.

Gibbs, M. 1946: *The Complete Adventures of Snugglepot and Cuddlepie*. Sydney: Angus and Robertson.

Gilbert, W. S. 1959: *The Savoy Operas*. London: Macmillan.

Glass, G. V. and Smith, M. L. 1978: *Meta-analysis of Research on the Relationship of Class Size and Achievement*. Boulder, Col.: University of Colorado.

Gold, M. 1975: *Try Another Way* (film). Indianapolis: Film Productions of Indianapolis.

Gordon, J. E. 1976: *The New Science of Strong Materials*, 2nd edn. Harmondsworth, England: Penguin.

Gordon, W. J. J. 1961: *Synectics: The Development of Creative Capacity*. New York: Harper and Row.

Gould, C. D. 1978: Practical work in sixth-form biology. *Journal of Biological Education*, 12, 33–8.

Greeno, J. G. 1973: The structure of memory and the process of solving problems. in R. L. Solso (ed.), *Contemporary Issues in Cognitive Psychology: The Loyola Symposium*. New York: Wiley.

de Groot, A. D. 1965: *Thought and Choice in Chess*. The Hague: Mouton.

Guesne, E., Tiberghien, A. and Delacôte, G. 1978: Méthodes et résultats concernant l'analyse des conceptions des élèves dans differents domaines de la physique. *Revue Français de Pédagogie*, 45, 25–32.

Gunstone, R. F. 1980: 'Structural Outcomes of Physics Instruction.' Unpublished doctoral thesis, Melbourne: Monash University.

Gunstone, R. F., Champagne, A. B. and Klopfer, L. E. 1981: Instruction for understanding: a case study. *Australian Science Teachers' Journal*, 27(3), 27–32.

Gunstone, R. F. and White, R. T. 1981: Understanding of gravity: *Science Education*, 65, 291–9.

Gunstone, R. F. and White, R. T. 1986: Assessing understanding by means of Venn diagrams. *Science Education*, 70, 151–8.

Haber-Schaim, U., Cross, J. B., Dodge, J. H. and Walter, J. A. 1971: *PSSC Physics*, 3rd edn. Lexington, Mass.: Heath.

Hamilton, N. R. 1964: Effects of logical versus random sequencing of items in an autoinstructional program under two conditions of covert response. *Journal of Educational Psychology*, 55, 258–66.

Hasan, O. E. and Billeh, V. Y. 1975: Relationship between teachers' change in attitudes toward science and some professional variables. *Journal of Research in Science Teaching*, 12, 247–53.

Hearnshaw, L. S. 1979: *Cyril Burt, Psychologist*. London: Hodder and Stoughton.

Helm, H. 1980: Misconceptions in physics amongst South African students. *Physics Education*, 15, 92–7, 105.

Helm, H. and Novak, J. D. (eds) 1983: *Proceedings of the International Seminar on Misconceptions in Science and Mathematics*. Ithaca, New York: Cornell University.

Hewson, M. G. A'B. 1982: 'Students' Existing Knowledge as a Factor Influencing the Acquisition of Scientific Knowledge.' Unpublished PhD thesis, University of the Witwatersrand, Johannesburg, South Africa.

Hilton, J. 1934: *Goodbye, Mr. Chips*. London, Hodder and Stoughton.

Hofman, H. H. 1977: An assessment of eight-year-old children's attitudes toward science. *School Science and Mathematics*, 77, 662–70.

Holley, C. D. and Dansereau, D. F. 1984: Networking: The technique and the empirical evidence. In C. D. Holley and D. F. Dansereau (eds), *Spatial Learning Strategies: Techniques, Applications, and Related Issues*. Orlando, Fla: Academic Press.

Holley, C. D., Dansereau, D. F., McDonald, B. A., Garland, J. C. and Collins, K. W. 1979: Evaluation of a hierarchical mapping technique as an aid to prose processing. *Contemporary Educational Psychology*, 4, 227–37.

Holliday, W. G. 1975: The effects of verbal and adjunct pictorial-verbal information in science instruction. *Journal of Research in Science Teaching*, 12, 77–83.

Hollowood, B. (ed.) 1963: *Pick of Punch*. London: Arthur Barker.

Holt, R. R. 1964: Imagery: the return of the ostracized. *American Psychologist*, 12, 254–64.

Huxley, A. 1932: *Brave New World*. London: Chatto and Windus.

Hynes, D. 1986: (untitled chapter) In J. R. Baird and I. J. Mitchell (eds), *Improving the Quality of Teaching and Learning: An Australian Case Study - The PEEL Project*. Melbourne: Monash University.

Jensen, A. R. 1969: How much can we boost IQ and scholastic achievement? *Harvard Educational Review*, 39, 1–123.

Johnson, P. E. 1964: Associative meaning of concepts in physics. *Journal of Educational Psychology*, 55, 84–8.

Johnstone, A. H. and El-Banna, H. 1986: Capacities, demands and processes – a predictive model for science education. *Education in Chemistry*, 23, 80–4.

Kamin, L. 1974: *The Science and Politics of IQ*. Potomac, Md: Erlbaum.

Kempa, R. F. and Dubé, G. E. 1974: Science interest and attitude traits in students subsequent to the study of chemistry at the ordinary level of the General Certificate of Education. *Journal of Research in Science Teaching*, 11, 361–70.

Kneller, G. F. 1941: *The Educational Philosophy of National Socialism*. New Haven, Conn.: Yale University Press.

Koestler, A. 1964: *The Act of Creation*. London: Hutchinson.

Kothandapani, V. 1971: Validation of feeling, belief, and intention to act as three components of attitude and their contribution to prediction of contraceptive behaviour. *Journal of Personality and Social Psychology*, 19, 321–33.

Kreitler, H. and Kreitler, S. 1974: The role of the experiment in science education. *Instructional Science*, 3, 75–88.

Krockover, G. H. and Malcolm, M. D. 1978: The effects of the Science Curriculum Improvement Study upon a child's attitude toward science. *School Science and Mathematics*, 78, 575–84.

Kruglak, H. 1953: Achievements of physics students with and without laboratory work. *American Journal of Physics*, 21, 14–16.

Krutetskii, V. A. 1976: *The Psychology of Mathematical Abilities in Schoolchildren* (J. Teller, trans.). Chicago: University of Chicago Press. (Original work published 1968.)

Lamb, W. G., Davis, P., Leflore, R., Hall, C., Griffin, J. and Holmes, R. 1979: The effect on student achievement of increasing kinetic structure of teachers' lectures. *Journal of Research in Science Teaching*, 16, 223–7.

Layton, D. 1973: *Science for the People: The Origins of the School Science Curriculum in England*. London: Allen and Unwin.

Lazarowitz, R., Barufaldi, P. J. and Huntsberger, P. J. 1978: Student teachers' characteristics and favorable attitudes toward inquiry. *Journal of Research in Science Teaching*, 15, 559–66.

Levin, G. R. and Baker, B. L. 1963: Item scrambling in a self-instructional program. *Journal of Educational Psychology*, 54, 138–43.

Likert, R. 1932: A technique for the measurement of attitudes. *Archives of Psychology* (whole issue, no. 140).

Loftus, E. F. 1979: *Eyewitness Testimony*. Cambridge, Mass.: Harvard University Press.

Lu, P. K. 1978: Three integrative models of kinetic structure in teaching astronomy. *Journal of Research in Science Teaching*, 15, 249–55.

Lundgren, U. P. 1977: *Model Analysis of Pedagogical Processes*. Stockholm: Institute of Education Department of Educational Research.

Luria, A. R. 1968: *The Mind of a Mnemonist* (L. Solotaroff, trans.). New York: Basic Books. (Original work published 1965.)

Lynch, P. and Strube, P. 1983: Tracing the origins and development of the modern science text: Are new text books really new? *Research in Science Education*, 13, 233–43.

MacKay, D. G. 1973: Aspects of the theory of comprehension, memory and attention. *Quarterly Journal of Experimental Psychology*, 25, 22–40.

McKellar, P. 1972: Imagery from the standpoint of introspection. In P. W. Sheehan (ed.), *The Function and Nature of Imagery*. New York: Academic Press.

Mackenzie, A. A. and White, R. T. 1982: Fieldwork in geography and long term memory structures. *American Educational Research Journal*, 19, 623–32 (copyright 1982, American Educational Research Association, Washington DC).

Markman, E. M. 1979: Realising that you don't understand: Elementary school children's awareness of inconsistencies. *Child Development*, 50, 643–55.

Marks, D. E. 1973: Visual imagery differences in the recall of pictures. *British Journal of Psychology*, 64, 17–24.

Maslow, A. 1970: *Motivation and Personality*, 2nd edn. New York: Harper and Row.

Mason, E. 1970: *Collaborative Learning*, London: Ward Lock.

Mathis, P. M. and Schrum, J. W. 1977: The effect of kinetic structure on achievement and total attendance time in audio-tutorial biology. *Journal of Research in Science Teaching*, 14, 105–15.

Mead, M. and Metraux, R. 1957: The image of the scientist among high school students: A pilot study. *Science*, 126, 384–90.

Mercurio, J. A. 1972: *Caning: Educational Ritual*. New York: Holt, Rinehart and Winston.

Meyer, B. J. F. 1975: *The Organization of Prose and its Effects on Memory*. Amsterdam: North-Holland.

Meyer, B. J. F., Brandt, P. M. and Bluth, G. J. 1980: Use of top-level structure in text: Key for reading comprehension in ninth-grade students. *Reading Research Quarterly*, 16, 72–103.

Miller, G. A. 1956: The magical number seven, plus or minus two: Some limits on our capacity for processing information. *Psychological Review*, 63, 81–97.

Moreira, M. A. 1980: A non-traditional approach to the evaluation of laboratory instruction in general physics courses. *European Journal of Science Education*, 2, 441–8.

Mori, I., Kitagawa, O. and Tadang, N. 1974: The effect of religious ideas on a child's conception of time: A comparison of Japanese children and Thai children. *Science Education*, 58, 519–22.

Munby, H. and Russell, T. 1983: A common curriculum for the natural sciences. In G. D. Fenstermacher and J. I. Goodlad (eds), *Individual Differences and the Common Curriculum*. 82nd Yearbook of the National Society for the Study of Education, Part 1. Chicago: University of Chicago Press.

Murray, H. A. 1938: *Explorations in Personality*. New York: Oxford University Press.

Neill, A. S. 1937: *That Dreadful School*. London: Herbert Jenkins.

Niedemeyer, F., Brown, J. and Sulzen, B. 1969: Learning and varying sequences of ninth-grade mathematics materials. *Journal of Experimental Education*, 37(3), 61–6.

Novak, J. D. and Gowin, D. B. 1984: *Learning How to Learn*. Cambridge: Cambridge University Press.

Nussbaum, J. 1979: Children's conceptions of the Earth as a cosmic body: a cross age study. *Science Education*, 63, 83–93.

Nussbaum, J. and Novak, J. D. 1976: An assessment of children's concepts of the Earth utilizing structured interviews. *Science Education*, 60, 535–50.

O'Brian, P. 1972: *Post Captain*. London: Collins.

Ormerod, M. B. and Wood, C. 1983: A comparative study of three methods of measuring the attitudes to science of 10- to 11-year-old pupils. *European Journal of Science Education*, 5, 77–86.

Ornstein, R. E. 1972: *The Psychology of Consciousness*. San Francisco: W. H. Freeman.

Orwell, G. 1945: *Animal Farm*. London: Secker and Warburg.

Osborne, R. J. 1976: Using student attitudes to modify instruction in physics. *Journal of Research in Science Teaching*, 13, 525–31.

Osborne, R. J. 1980: *Force*. Hamilton, NZ: University of Waikato, Learning in Science Project, paper no. 16.

Osborne, R. J. and Freyberg, P. 1985: *Learning in Science: The Implications of Children's Science*. Auckland, NZ: Heinemann.

Osborne, R. J. and Gilbert, J. K. 1980: A method for investigating concept understanding in science. *European Journal of Science Education*, 2, 311–21.

Osgood, C. E., Suci, G. J. and Tannenbaum, P. H. 1957: *The Measurement of Meaning*. Urbana, Ill.: University of Illinois Press.

Paivio, A. 1969: Mental imagery in associative learning and memory. *Psychological Review*, 76, 241–63.

Paivio, A. 1971: *Imagery and Verbal Processes*. New York: Holt, Rinehart and Winston.

Paris, S. G. and Jacobs, J. E. 1984: The benefits of informed instruction for children's reading awareness and comprehension skills. *Child Development*, 53, 2083–93.

Payne, D. A., Krathwohl, D. R. and Gordon, J. 1967: The effect of sequence upon programmed instruction. *American Educational Research Journal*, 4, 125–32.

Physical Science Study Committee (PSSC) 1965: *Physics*, 2nd edn. Boston, Mass.: Heath.

Piaget, J. 1953: *Logic and Psychology*. (W. Mays and F. Whitehead, trans.). Manchester: Manchester University Press.

Posner, G. J., Strike, K. A., Hewson, P. W. and Gerzog, W. A. 1982: Accommodation of a scientific conception: Toward a theory of conceptual change. *Science Education*, 66, 211–27.

Poster, C. 1982: *Community Education: Its Development and Management*. London: Heinemann.

Preece, P. F. W. 1976a: Associative structure of science concepts. *British Journal of Educational Psychology*, 46, 174–83.

Preece, P. F. W. 1976b: Mapping cognitive structure: a comparison of methods. *Journal of Educational Psychology*, 68, 1–8.

Project Physics Course, 1970: New York: Holt, Rinehart and Winston.

Pylyshyn, Z. W. 1973: What the mind's eye tells the mind's brain. *Psychological Bulletin*, 80, 1–24.

Redfield, D. L. and Rousseau, E. W. 1981: A meta-analysis of experimental research on teacher questioning behavior. *Review of Educational Research*, 51, 237–45.

Rokeach, M. 1970: *Beliefs, Attitudes, and Values*. San Francisco: Jossey-Bass.

Rowe, M. B. 1969: Science, silence and sanctions. *Science and Children*, 6(6), 11–13.

Rowe, M. B. 1974a: Wait-time and rewards as instructional variables, their influence on language, logic and fate control: Part 1. Wait time. *Journal of Research in Science Teaching*, 11, 81–94.

Rowe, M. B. 1974b: Relation of wait-time and rewards to the development of language, logic, and fate control: Part 2. Rewards. *Journal of Research in Science Teaching*, 11, 291–308.

Ryle, G. 1949: *The Concept of Mind*. London: Hutchinson.

Sassoon, S. 1928: *Memoirs of a Fox-Hunting Man*. London: Faber and Gwyer.

Schank, R. C. and Abelson, R. P. 1977: *Scripts, Plans, Goals and Understanding: An Inquiry into Human Knowledge Structures*. Hillsdale, NJ: Erlbaum.

Selmes, C. 1973: Nuffield A-level biology: Attitudes to science. *Journal of Biological Education*, 7(4), 43–7.

Shavelson, R. J. 1972: Some aspects of the correspondence between content structure and cognitive structure in physics instruction. *Journal of Educational Psychology*, 63, 225–34.

Shavelson, R. J. 1973: Learning from physics instruction. *Journal of Research in Science Teaching*, 10, 101–11.

Shavelson, R. J. 1974: Methods for examining representations of a subject-matter structure in a student's memory. *Journal of Research in Science Teaching*, 11, 231–49.

Shavelson, R. J. and Stanton, G. C. 1975: Construct validation: Methodology and application to three measures of cognitive structure. *Journal of Educational Measurement*, 12, 67–85.

Shayer, M. and Adey, P. 1981: *Towards a Science of Science Teaching: Cognitive Development and Curriculum Demand*. London: Heinemann.

Sherwood, R. D. and Herron, J. D. 1976: Effect on student attitude: Individualized IAC versus conventional high school chemistry. *Science Education*, 60, 471–4.

Shulman, L. S. and Tamir, P. 1973: Research on teaching in the natural sciences. In R. M. W. Travers (ed.), *Handbook of Research on Teaching*, 2nd edn. Chicago: Rand McNally.

Simmons, E. S. 1980: The influence of kinetic structure in films on biology students' achievement and attitude. *Journal of Research in Science Teaching*, 17, 67–73.

Skinner, B. F. 1968: *The Technology of Teaching*. New York: Appleton Century Crofts.

Smail, B. and Kelly, A. 1984: Sex differences in science and technology among 11-year-old schoolchildren: II-affective. *Research in Science and Technological Education*, 2, 87–106.

Stavy, R. and Berkowitz, B. 1980: Cognitive conflict as a basis for teaching quantitative aspects of the concept of temperature. *Science Education*, 64, 679–92.

Stead, B. F. and Osborne, R. J. 1980: Exploring science students' concepts of light. *Australian Science Teachers Journal*, 26(3), 84–90.

Stead, K. and Osborne, R. J. 1981: *Friction*. Hamilton, NZ: University of Waikato, Learning in Science Project, paper no. 19.

Stewart, J. 1979: Content and cognitive structure: Critique of assessment and representation techniques used by science education researchers. *Science Education*, 63, 395–405.

Street, R. F. 1931: *A Gestalt Completion Test*. New York: Teachers College, Columbia, Bureau of Publications.

Strike, K. A. and Posner, G. J. 1976: Epistemiological perspectives on conceptions of curriculum organization and learning. *Review of Research in Education*, 4, 106–41.

Suchman, J. R. 1966: *Inquiry Development Program: Developing Inquiry*. Chicago: Science Research Associates.

Sutton, C. R. 1980: The learner's prior knowledge: a critical review of techniques for probing its organization. *European Journal of Science Education*, 2, 107–20.

Symington, D. J. and White, R. T. 1983: Children's explanations of natural phenomena. *Research in Science Education*, 13, 73–81.

Tasker, R. 1981: Children's views and classroom experiences. *Australian Science Teachers Journal*, 27(3), 33–7.

Theobald, J. H. 1980: The interaction of student attributes and teaching style. *Journal of Biological Education*, 14(3), 231–6.

Thesiger, W. 1959: *Arabian Sands*. New York: Dutton.

Thurber, J. 1933: *My Life and Hard Times*. New York: Harper and Brothers.

Thurstone, L. L. 1928: Attitudes can be measured. *American Journal of Sociology*, 33, 529–54.

Tiberghien, A. and Delacôte, G. 1976: Manipulation et représentations des circuits électriques simple par des enfants de 7 a 12 ans. *Revue Française de Pédagogie*, 34, 32.

Tobin, K. G. 1980: The effect of an extended teacher wait-time on science achievement. *Journal of Research in Science Teaching*, 17, 469–75.

Tolman, R. R. and Barufaldi, J. P. 1979: The effects of teaching the Biological Sciences Curriculum Study Elementary School Sciences Program on attitudes toward science among elementary school teachers. *Journal of Research in Science Teaching*, 16, 401–6.

Trembath, R. J. and White, R. T. 1979: Mastery achievement of intellectual skills. *Journal of Experimental Education*, 47, 247–52.

Tulving, E. 1972: Episodic and semantic memory. In E. W. Tulving and W. Donaldson (eds), *Organization of Memory*. New York: Academic Press.

Wagner, R. K. and Sternberg, R. J. 1984: Alternative conceptions of intelligence and their implications for education. *Review of Educational Research*, 54, 179–223.

Weinstein, C. F. and Mayer, R. E. 1986: The teaching of learning strategies. In M. C. Wittrock (ed.), *Handbook of Research on Teaching*, 3rd edn. New York: Macmillan.

West, L. H. T. and Pines, A. L. (eds) 1985: *Cognitive Structure and Conceptual Change*. Orlando, Fl: Academic Press.

White, R. T. 1982: Memory for personal events. *Human Learning*, 1, 171–83.

White, R. T. and Gunstone, R. F. 1980: *Converting Memory Protocols to Scores on Several Dimensions*. Australian Association for Research in Education Annual Conference Papers, 486–93.

White, R. T. and Tisher, R. P. 1986: Research on natural sciences. In M. C. Wittrock (ed.), *Handbook of Research on Teaching*, 3rd edn. New York: Macmillan.

The White House Transcripts 1973. Chronology by Linda Amster. General editor: Gerald Gold. Editorial note copyright © 1973, 1974 by The New York Times Company. Reprinted by permission of Bantam Books. All rights reserved.

Whorf, B. L. 1940: Science and linguistics. In J. B. Carroll (ed.), *Language, Thought, and Reality*. Cambridge, Mass.: MIT Press.

Winne, P. H. 1979: Experiments relating teachers' use of higher cognitive questions to student achievement. *Review of Educational Research*, 49, 13–50.

Wintle, J. and Kenin, R. (eds) 1978: *The Dictionary of Biographical Quotation of British and American Subjects*. London: Routledge and Kegan Paul.

Wong, B. Y. L. and Jones, W. 1982: Increasing metacomprehension in learning-disabled and normally-achieving students through self-questioning training. *Learning Disability Quarterly*, 5, 228–40.

Yager, R. E., Engen, H. B. and Snider, B. C. 1969: Effects of the laboratory and demonstration methods upon the outcomes of instruction in secondary biology. *Journal of Research in Science Teaching*, 6, 76–86.

Yates, F. A. 1966: *The Art of Memory*. London: Routledge and Kegan Paul.

Index